Clinical Evaluation of Medical Devices

Clinical Evaluation of Medical Devices

Principles and Case Studies

Edited by

Karen Becker Witkin

THE WEINBERG GROUP, INC.
Washington, DC

Humana Press ✳ **Totowa, New Jersey**

All authored papers, comments, opinions, conclusions, or recommendations are those of the author(s), and do not necessarily reflect the views of the publisher.

For additional copies, pricing for bulk purchases, and/or information about other Humana titles, contact Humana at the above address or at any of the following numbers: Tel: 201-256-1699; Fax: 201-256-8341; E-mail: humana@mindspring.com, or visit our Website: http://humanapress.com

This publication is printed on acid-free paper. ∞
ANSI Z39.48-1984 (American Standards Institute) Permanence of Paper for Printed Library Materials.

Cover design by Patricia F. Cleary.

Cover illustrations *(clockwise from top left)*: Fig. 1 in Chapter 12, "Polyurethane Pacemaker Leads: *The Contribution of Clinical Experience to the Elucidation of Failure Modes and Biodegradation Mechanisms,"* by Ken Stokes; Fig. 2 in Chapter 6, "Prospective Multicenter Clinical Trials in Orthopedics: *Special Concerns and Challenges,"* by John D. Van Vleet; Fig. 3 in Chapter 11, "Role of Device Retrieval and Analysis in the Evaluation of Substitute Heart Valves," by Frederick J. Schoen; Fig. 5 in Chapter 12; Fig. 1B in Chapter 11; and Fig. 4 in Chapter 12.

Photocopy Authorization Policy:

Printed in the United States of America. 10 9 8 7 6 5 4 3 2 1

Library of Congress Cataloging in Publication Data

Main entry under title:

Clinical evaluation of medical devices/edited by Karen Becker Witkin.
      p.   cm.
   Includes index.
   ISBN 0-89603-446-1 (alk. paper)
   1. Medical instruments and apparatus—Evaluation. 2. Clinical trials. I. Witkin, Karen Becker
   [DNLM: 1. Equipment and Supplies—standards. 2. Device Approval—standards. 3. Clinical Trials—standards. 4. Evaluation Studies. W 26 C641   1998
   R856.C545   1998
   610'.28—dc21
   DNLM/DLC
   for Library of Congress   97-23205
                                                      CIP

Foreword

The world is changing rapidly, and nowhere is this more apparent than in medicine. The standards are rapidly rising in the field of medical device trials. A few years ago, device developers would look askance if one told them that medical device trials and drug trials should have the same standards. Today, such a statement does not seem as outrageous, although there is still a large gap in the design of trials and number of trials conducted for medical device and drug development programs. More than 20 years after the enactment of the US Medical Device Amendments, we can see that they served as an impetus to raise clinical trial standards for devices.

Whether the data to establish the safety and efficacy of a device come from one, two, or even more clinical trials is less important in evaluating the device than whether the data are medically and scientifically supportive of its safety and efficacy. Having at least two separate studies, and at least two sites confirm results, adds a great deal of scientific credibility and support to a conclusion of safety and efficacy, even though a confirmatory trial is not yet a regulatory requirement in most countries.

It used to be easy to identify a device and a drug, but distinctions between the two are becoming blurred. There are devices that are pure chemicals and others that are taken internally. Several drug–device combinations are already available or are currently in development. Applying the clinical standards for devices to such combination trials raises several questions that must be addressed before such clinical trials are initiated (for example, determining whether a drug–device combination should adhere to either the drug or the device standards).

Many people claim that device trials need not be held to high clinical standards because it is sometimes unethical to conduct such controlled trials. Although this is occasionally true (trials may also be unethical for some drugs), it is important not to use this as an excuse to avoid designing controlled trials whenever possible. Unfortunately, it is often the regulators who serve as enforcers, ensuring that this approach is followed. *Clinical Evaluation of Medical Devices: Principles and Case Studies* presents clear discussions of appropriate trial designs that may be used when the most rigorous designs are not possible. The readers of this book are indeed fortunate to have so many talented authors writing about this topic.

Given the enormous heterogeneity of medical devices, the case studies in Part II presenting valuable real-life experiences should provide great benefit to readers. Although the reader working in one particular area may focus on the closest parallel chapter in Part II, there are numerous reasons why all chapters should be skimmed, if not read closely, for useful information. A careful reader will benefit immeasurably from the many experiences and lessons shared in these pages. I believe that the dedication of two-thirds of the chapters to case studies is a noteworthy strength of this book.

Karen Becker Witkin has assembled an outstanding group of authors and has organized *Clinical Evaluation of Medical Devices: Principles and Case Studies* in a manner that will make sense to both experts and tyros in the device industry. Many device companies are struggling with issues of developing their products as rapidly as possible, as well as the regularly escalating standards for clinical trials. This book provides a state-of-the-art overview that will enable companies to develop their products more efficiently.

Adhering to the highest possible standards in pivotal clinical studies makes good business sense because it will generate reliable data and will enable regulatory authorities to review and approve a dossier more rapidly. This enables the corporate marketing group to provide convincing data to physicians, who are the ultimate customers for the devices, because claims and performance statements can then be made for their devices that are better than those of their competitors. Another benefit of adhering to high standards is that third party payors and formulary groups can also be given convincing data. Moreover, if the product does not meet the minimally acceptable criteria established in advance of the trial, it then allows the company to terminate the project with the least psychic and monetary expense. This can be an important factor in helping a company move on to its next project, avoiding wasted time and energy on a project of little or no value.

Bert Spilker, MD, PhD
President, Orphan Medical, Inc.
Minnetonka, MN

Preface

The medical device industry is renowned for its innovative products. The successful conception, development, and continuous improvement of thousands of novel medical devices and health care technologies over the last 50 years is a testament to the productive collaboration between engineers and physicians fostered by this industry. Today, the continuing evolution in the design of new medical devices coincides with the evolution of techniques for testing these products—both during development and post-marketing. In a health care environment that demands evidence-based medicine, clinical evaluation of medical devices is increasingly a focal point of systematic research, particularly for implanted devices and technologies that significantly impact on patient health.

Clinical Evaluation of Medical Devices: Principles and Case Studies is an attempt to capture the foundations and best experience of this clinical research area, now maturing into a specialty science with distinct methodologies and objectives. In recent years, the singular research approaches required for clinical studies of device performance have been enhanced by concerted efforts to adapt and incorporate the principles of rigorous clinical research developed for pharmaceutical products. Part I of this work provides an overview of the role of clinical research in the development and marketing of medical devices, the basic principles used in the design of such studies, and international regulatory requirements for research and registration of medical devices. Part II is a series of case studies selected to illustrate the broad array of study designs that have been successfully applied to many different research problems and a variety of therapeutic or diagnostic products.

Clinical Evaluation of Medical Devices: Principles and Case Studies evolved from a course I organized in collaboration with the US Food and Drug Administration, generously sponsored by the Society for Biomaterials. I am grateful to those colleagues who graciously contributed their time and expertise in the form of chapters and ideas contributing to the final concept. Additionally, without the expert and enthusiastic assistance of my associates at THE WEINBERG GROUP INC., particularly Diane Mandell and Diane Wallerson, this book would not have been possible. Finally, I am indebted to Thomas O. Henteleff and Daniel R. Dwyer (Kleinfeld, Kaplan, and Becker) for introducing me to the complex and fascinating world of medical device regulation, and to Myron S. Weinberg and Matthew R. Weinberg for their substantial and unquestioning support to the pursuit of my interest in this field.

Karen Becker Witkin
THE WEINBERG GROUP, INC.
Washington, DC

Contents

Contributors

FRANK DeLUSTRO, PhD • *Cohesion Inc., Palo Alto, CA*

FREDERICK J. DOREY, PhD • *Department of Orthopaedic Surgery and Statistics, UCLA School of Medicine, Los Angeles, CA*

FRANK C. DORSEY, PhD • *Biometric Research Institute, Inc., Arlington, VA*

ROSS ERICKSON, PhD • *Collagen Corporation, Palo Alto, CA*

ROBERTO FIORENTINI, MD • *FIDIA Pharmaceutical Corp., Washington, DC*

MICHELE GARGANO, PhD • *Kunitz and Associates, Inc., Rockville, MD*

GARY L. GRUNKEMEIER, PhD • *St. Vincent Heart Institute, Portland, OR*

RENE KOZLOFF, PhD • *Kunitz and Associates, Inc., Rockville, MD*

SELMA A. KUNITZ, PhD • *Kunitz and Associates, Inc., Rockville, MD*

DIANE E. MANDELL, PhD • *THE WEINBERG GROUP, INC., Washington, DC*

ROSANNE B. McTYRE, PhD • *THE WEINBERG GROUP, INC., Washington, DC*

ROLAND MOSKOWITZ, MD • *Division of Rheumatic Diseases, Department of Medicine, University Hospitals of Cleveland, OH*

WAYNE R. PATTERSON, PhD • *Department of Pathology, The University of Texas Medical Branch, Galveston, TX*

LINDA M. POTTERN, PhD • *Office of Disease Prevention, National Institutes of Health, Bethesda, MD*

FREDERICK J. SCHOEN, MD, PhD • *Department of Pathology, Brigham and Women's Hospital, Boston, MA*

SHARON A. SEGAL, PhD • *THE WEINBERG GROUP, INC., Washington, DC*

BERT SPILKER, MD, PhD • *Orphan Medical, Minnetonka, MN*

KEN STOKES, PhD • *Medtronic, Inc., Minneapolis, MN*

JOHN D. VAN VLEET, PhD • *DePuy, Inc., Warsaw, IN*

KAREN BECKER WITKIN, PhD • *THE WEINBERG GROUP, INC., Washington, DC*

Part I

**Fundamentals of Clinical Study
Design and Evaluation**

1

Clinical Trials in Development and Marketing of Medical Devices

Karen Becker Witkin

1. Introduction

Medical devices are health-care products distinguished from drugs for regulatory purposes in most countries throughout the world based on mechanism of action. Unlike drugs, medical devices operate via physical or mechanical means and are not dependent on metabolism to accomplish their primary intended effect. As defined in the US Food, Drug and Cosmetic Act, the term medical device

> "...means an instrument, apparatus, implement, machine, contrivance, implant, *in vitro* reagent, or other similar or related article... intended for use in the diagnosis of disease or other conditions, or in the cure, mitigation, treatment, or prevention of disease...or intended to affect the structure or any function of the body...and which does not achieve its primary intended purposes through chemical action within or on the body...."[1]

This broad definition of medical devices encompasses literally tens of thousands of different types of health-care products, including in vitro diagnostics.

Developing new medical devices and extending the scope of what is known about the performance of already marketed products often requires clinical investigations. Developing well-controlled prospective clinical trials of medical devices presents design challenges that

From: *Clinical Evaluation of Medical Devices: Principles and Case Studies*
Edited by K. B. Witkin Humana Press Inc., Totowa, NJ

are unique from those faced in studies of pharmaceuticals. For example, clinical outcomes observed in medical-device studies are influenced not only by the product under evaluation and the patient, but also by the skill and discretion of the user. The user is most often a health-care professional but is sometimes the patient. The impact of this third parameter—the medical device user—is a variable unique to medical device studies and can be responsible for the greatest degree of variability in the clinical outcomes measured. Being aware of, and controlling for, the influence of the user on device performance is a critical variable that requires attention in designing a clinical study. Other critical features typically considered in the design of well-controlled studies, such as the choice of a control group, the need to reduce bias, and the need to control for confounders, are common to both drug and device trials. However, the nature of the difficulties presented and the approaches used to successfully address these difficulties are often different.

This chapter provides a detailed discussion of the features of medical devices that can pose challenges in the design of well-controlled clinical studies and methods for addressing these design challenges. An overview of the role of clinical research in the life-cycle of medical device product development and marketing, and the essential elements of a clinical investigational plan for a prospective medical device clinical trial are also presented.

2. Drugs Versus Devices: Is There a Difference?

The Medical Devices Amendments to the Federal Food, Drug, and Cosmetic Act signed into law in 1976 provided the Food and Drug Administration (FDA) with broad jurisdiction and authority over the commercialization of medical devices. Prior to the development of that legislation, the Secretary of Health and Human Services assembled a task force to consider the nature of the medical device industry in the United States, the extent to which the products of this industry should be subject to regulation, and the best mechanisms for protecting the public health without applying an undue burden to industry or preventing innovation. The task force, commonly called the Cooper Committee after Dr. Theodore Cooper (at that time, Director of the National Heart, Lung, and Blood Institute), submitted a report in 1970, the conclusions of which formed the framework for the legislation passed in 1976.[2]

The Cooper Committee concluded that medical devices are unlike pharmaceuticals in significant ways, and as such, direct application of the "drug-model" of regulation to these products is not desirable or feasible. Instead, a novel regulatory approach, based on the extent of risk posed to the patient from the use of the device, was recommended. Among the "...inherent differences between drugs and devices..."[2] noted by the Cooper Committee are that medical devices are an extremely diverse group of products varying widely in their intended uses and principles of operation, they are often designed by physicians and subject to frequent innovation in both design and use, they are used primarily by health-care professionals rather than patients, they are most often developed by small companies, and on a per product basis their annual sales are only a fraction of that for a typical pharmaceutical product.

Based on the Cooper Committee recommendations and subsequent testimony, Congress developed legislation incorporating a regulatory pathway for medical devices based on the consideration of the relative risk posed by each product and an apparent acceptance of the principle that, in spite of and beyond any Federal regulatory requirements, the skill and clinical judgement of the user ultimately play a major role in the performance of any particular medical device. The premarket approval regulations for medical devices incorporate a requirement that valid scientific evidence of safety and effectiveness be provided by sponsors, but unlike for drugs, this evidence can come from sources other than well-controlled clinical investigations, such as "...partially controlled studies, studies and objective trials without matched controls, well-documented case histories conducted by qualified experts, and reports of significant human experience with a marketed device."[3] However, the regulation specifically precludes reports of clinical experience that are not adequately supported by data, such as anecdotal reports or opinion.

The standard for approval of medical devices is also more flexible than for drugs in that the regulations require "reasonable assurance" of safety and effectiveness, rather than the more onerous burden of "substantial evidence" specified for drugs.[4,5] Although FDA's Center for Devices and Radiological Health has taken responsible and aggressive steps to ensure the rigor of clinical research required to support registration of new medical devices, the differential standard of evidence for approval of devices vs drugs has been recently reaffirmed by the Agency.[6, 7] As noted by the FDA, the primary practical consequence of the regulations is that the approval requirements for drugs require replication of clinical findings (i.e., more than one clinical trial), but for devices a single pivotal clinical trial is sufficient for approval since "...for medical devices, where the mechanism of action is a result of product design and substantially verified by in vitro performance testing, the agency has routinely relied on single studies evaluated for internal and across-center consistency."*

The differences between drugs and medical devices identified by the Cooper Committee 25 years ago remain valid today, and some of these differences are key to understanding important features of medical devices that influence product development plans and clinical study design (*see* Table 1).

2.1. Devices Are Primarily Used by Health Care Professionals

In studies of pharmaceuticals, the two principal variables that interact are the drug and the patient. Given that the investigators are able to control for other variables, such as concomitant exposures, pre-existing conditions, and the progression of disease, the outcomes measured are ultimately a function of the interaction of the drug and the patient. In contrast, the interaction of a medical device and the patient is usually mediated by a responsible third party, the user of the product, most often a health-care professional or surgeon. Thus, the clinical outcomes measured in the study of safety and effectiveness for a medical device are a function not only of the interaction between the device and the patient, but also of the skill of the user. The user as an intermediary poses two major difficulties in the design of a clinical trial for devices that are not commonly encountered in studies of pharmaceuticals: Users can rarely be blinded to the treatment intervention and users themselves can have an impact on the performance of the product. Indeed, the user is an integral variable in the performance of the product. A device performs better in the hands of an experienced user than in the hands of a naive user, a phenomenon typically called the "learning curve." Training in the use of a device is an integral part of the investigation of its clinical performance and eventual marketing. Variability in the proficiency

---

* The reader is referred to Chapter 4, for details on the regulatory requirements for registration and marketing of medical devices in the United States and the European Economic Community.

Table 1
Characteristics of Medical Devices that Impact on Clinical Trial Design

Characteristic	Clinical study design issue
Devices are primarily used by health care professionals.	Product performance is influenced by user. The user often cannot be blinded to the study intervention.
Devices are subject to frequent incremental innovation.	Bench testing and animal models alone may validate new designs. Ethical considerations may preclude comparative trials. Results from long-term clinical studies may no longer be relevant to current products and medical procedures.
Some devices are implanted.	Exposure to the product is not readily terminated, or without irreversible consequences to patient. Placebo control groups (sham-surgery) are not possible. Medically appropriate alternative treatment regimens may not be available to provide randomized, concurrent controls. Long-term performance evaluations primarily rely on design controls and failure analysis.
Devices are often developed by small companies; sales on a per-product basis are less than that for average pharmaceuticals.	Practical considerations (regulations, financial constraints) limit new product development and testing.

of the user, if not accurately assessed or minimized in importance, can lead to inaccurate estimates of device performance.

The fact that the user, and often the patient, cannot be blinded to the intervention under study can introduce bias into the assessment of clinical performance if the clinical investigator is jointly responsible for treatment and assessment of performance. For this reason, wherever possible, blinded evaluators are preferred to clinical investigators in assessing efficacy.

2.2. Devices Are Subject to Frequent, Incremental Innovations

Frequent innovations in the design and use of medical devices are standard practice in the industry. These are often minor modifications that enhance safety, reliability, patient comfort, or ease of use, and in most cases do not require regulatory approval or premarket notification. Bench testing and/or evaluations in animal models are often sufficient to validate the suitability of a design change without the need for controlled clinical trials. It is not uncommon for the results of in vitro performance evaluations on a new design to be sufficiently compelling that clinicians and Institutional Review Boards (IRBs) are reluctant to proceed with a comparative clinical trial, since continued use of the older design is deemed to be unethical. However, for design innovations intended to significantly improve performance parameters (efficacy) or to expand indications for use, clinical studies are usually necessary.

2.3. Some Devices Are Implanted

It is estimated that 6 million people in the United States alone have some type of implanted medical device, most commonly a fracture fixation product or an intraocular lens.[8] Whether for short-term or long-term use, implanted devices always result in irreversible effects on the patient, even if it is only the presence of residual scar tissue. Unlike a clinical experiment with drugs, exposure to an implanted medical device is not readily terminated. Clinical studies of implanted devices are surgical trials, a situation that precludes the use of placebo or sham-operated control groups. An ethically appropriate alternative treatment group may be difficult to identify, requiring the use of historical controls in the trial or patients as their own controls (pre- and post-surgery) to evaluate outcomes. Controlled, prospective, long-term performance evaluations of implanted devices (> 2 yr) are rare because of logistical constraints (i.e., large sample sizes are required to mitigate against loss to follow-up; inability to identify a sufficient number of suitable patients; expense of a large trial in relation to market size). Instead, information to track rare complications, identify failure modes, and contribute to enhanced designs is most often collected from analysis of case series, failure analysis of retrieved devices, bench testing, and formal design reviews.

2.4. Device Manufacturers May Be Small Companies

Medical devices are often products developed by small companies that generate annual sales revenue only a fraction of that generated by the average pharmaceutical. A responsible manufacturer conducts whatever testing is required to develop a safe and reliable product,

Table 2
Typical Classification of Medical Device Clinical Studies

Pilot studies of safety, performance, and/or design prior to marketing

Pivotal trials of safety and effectiveness prior to marketing

Postmarketing studies
    In support of expanded labeling claims
    In support of comparative performance claims
    Pharmacoeconomic studies
    Observational or analytical studies of specific safety or performance issues
    Explant retrieval and failure analysis investigations

---

but practical constraints experienced by this particular segment of the health-care industry are particularly influential in driving product development and testing decisions.

3. The Role of Clinical Studies in Product Development

Most commonly, the impetus for conducting a clinical study is to demonstrate the safety and effectiveness of an investigational device prior to marketing, a requirement for implants and other significant risk devices* when seeking registration in the United States and most international markets.[9] However, carefully conceived clinical research also has a role to play in enhancing product development and marketing for nonsignificant risk products, despite the fact that the vast majority of devices and diagnostics reach the market based solely on safety and performance testing in animal models and in vitro. Postmarketing studies can yield information to enhance product design, extend labeling claims, and provide data on comparative effectiveness and support for cost–benefit claims. A classification scheme for clinical investigations typically undertaken on medical devices is summarized in Table 2.

3.1. Pilot Studies

Pilot studies, or feasibility studies, are usually single-center studies in a limited number of patients designed to accomplish any number of objectives within a clinical testing program. Pilot studies are not usually designed as hypothesis-testing studies, but rather are meant

---

* A significant risk device is a product that presents a potential for serious risk to the health, safety, or welfare of a subject, and is most commonly an implant or life-sustaining product.

to generate data in support of the design of rigorous analytical (i.e., hypothesis-testing) trials. The first study of a novel investigational device in humans is usually a small pilot study undertaken to evaluate safety under carefully controlled conditions and to provide data in support of broader testing of performance in a larger population. Pilot studies are the first opportunity to evaluate the role of the user in device performance under actual clinical conditions and to gather information on design features that may be modified to optimize proper use of the device. Prior to designing a pivotal clinical trial to evaluate device safety and effectiveness, pilot studies allow the sponsor to collect data on a series of patient outcomes that may be related to device performance, thereby contributing to the identification and selection of clinically significant measures for use as effectiveness endpoints in a subsequent pivotal trial. Many times, the selection of measures of safety and effectiveness requires that validated methods for assessment be developed. More extensive pilot studies can incorporate validation of assessment tools and can be used to generate enough data on the interpatient variability of endpoints to support sample size calculations for use in the design of a hypothesis testing study.

3.2. Pivotal Trials of Safety and Effectiveness

A single, well-controlled clinical trial of device performance remains sufficient for approval of a significant risk device by the FDA. These are prospective, analytical studies that provide objective evidence of effectiveness based on single, or in some cases multiple, clinical outcomes of significance. In combination with data from bench testing and animal studies, results from a single trial are adequate to establish "reasonable evidence" of safety and effectiveness. When direct comparisons are made to alternative treatment options, effectiveness of the new device is expected to be *not worse than* that of other available devices or treatment. With rare exceptions, pivotal trials in support of successful FDA premarket approval applications (PMAs) are multicenter.

Clinical research conducted on an investigational device prior to marketing creates the foundation for claims that will appear on the label once marketing authorization is accomplished. This point is especially critical in the United States, where the expectations for reliable data in support of all aspects of the label are the most rigorous. For this reason, the clinical portion of the product development plan should never be considered in isolation from the ultimate marketing goal. In some cases, bench testing and animal studies can provide additional performance data to augment the clinical research and support expanded label claims.

3.3. Postmarketing Studies

Two categories of postmarketing studies can be distinguished: mandated postapproval studies and postmarket surveillance studies. It has become increasingly common for the FDA to require sponsors of Class III devices to conduct a "postapproval" clinical study as a mandatory condition of PMA approval. These studies are usually narrow in scope and focus on generating additional data to expand on results of pivotal trial(s) in support of product approval. The objectives of postapproval studies, whether mandated by a regulatory agency or the state of the science, typically include longer-term follow-up, additional data on the incidence and time-course to appearance of adverse events, and additional data in support of broader label claims (indications for use, duration of effectiveness, or product benefits). Postapproval studies may be undertaken as an extension of a pivotal trial via protocol amendments to extend follow-up or they may be conceived as an entirely separate study. The trend toward mandatory postapproval studies reflects a commitment by the FDA to work with sponsors whose investigational plans were finalized prior to 1993, when the Agency shifted to a more rigorous standard of clinical trial requirements.[6, 10, 11] Careful consideration of pivotal trial design and good communication with the FDA in the design of investigational plans prior to initiation of pivotal clinical trials will likely minimize the need for mandatory postapproval studies.

Distinct from postapproval studies are various types of postmarketing studies undertaken by a manufacturer to accomplish a variety of goals. These may be sponsor-initiated or may be required by a regulatory agency, for example, in response to a perceived safety concern. The US 1990 Safe Medical Devices Act empowered the FDA to require Mandatory Postmarket Surveillance Trials for certain types of devices, and Discretionary Postmarket Surveillance Trials when there is a perceived public health need. As of 1996, only a few Discretionary Postmarket Surveillance Trials had been requested by the FDA. As may be expected, all of these postmarket surveillance trials were directed at developing systematic data on either long-term failure

modes and/or the potential for serious adverse events occurring in a
small number of patients (heart valves, injectable collagen, polyurethane breast implants, pacemaker leads). It is worth remembering
that compliance with good manufacturing practices for medical devices
also compels sponsors to engage in postmarket surveillance monitoring of marketed products. This includes requirements to evaluate
and, if appropriate, act on complaints, product failures, and adverse
events associated with product use. This is not a passive process; the
responsible manufacturer maintains routine procedures for systematic evaluation of postmarket experience, directed toward investigating product failures and successes, and may include research (bench
testing or clinical) to improve product performance and safety.

Outside of regulatory requirements, other goals of postmarket
research can include performing comparative studies with alternative
or competitive treatments and/or devices aimed at providing support
for pharmacoeconomic claims, comparative effectiveness claims, or
expanded label claims. As previously noted, for implanted devices,
studies in which patients are followed prospectively for >2 yr are
not generally practical because of the loss of follow-up, enormous expense, and rapidly progressing changes in medical practice. Carefully
considered programs to exploit explant analysis and failure investigations, coupled with design controls prior to marketing, are the most
common, effective, and efficient means of gathering data on long-
term performance. Observational studies are also beginning to be
used effectively (*see* Chapter 2) for retrospective studies of clinical
experience, especially when the goal is to gather data on patient or
device-specific factors that may contribute to long-term performance
failures. As with pharmaceuticals, vigilance in the evaluation of
adverse events reports, returned goods, and complaints are important
sources of information on clinical experience. Although anecdotal,
these data are the foundation for product research into design deficiencies and strengths and lead to products that perform better. Examples of some successful postmarket investigations appear in Chapters
8, 11, and 12.

4. Elements of a Clinical Investigational Plan

Clinical research studies can be categorized as either observational
or analytical. An observational study is designed to collect and analyze
data without control by the investigator in terms of subject exposure

or treatment interventions. The data collected is recorded and analyzed in order to generate or test a hypothesis. Examples of observational studies that are designed to collect data include daily recording of rainfall and temperature, dietary intake surveys, and data collected on national cancer incidence. In cases in which an experimental approach is not practical or is otherwise unfeasible, observational studies are valuable. The special utility of observational studies in the study of medical device performance and safety is described in detail in Chapter 2.

Analytical studies, the subject of the discussion that follows, are hypothesis-testing trials comprised of a cohort exposed to a specific intervention, the impact of which is subsequently assessed. The "gold standard" of analytical clinical trial design is the prospective, randomized study with concurrent control group(s). The reader is referred to several excellent sources for detailed discussions of the principles of good clinical study design, conduct, and analysis. Although generally written from the point of view of pharmaceuticals, the principles of good clinical study design are equally applicable to both drugs and devices.[12-16]

Clinical protocols for medical device trials typically incorporate a device description and a patient risk analysis, in addition to other information necessary to describe the design, conduct, and analysis of the planned trial. The essential elements of a clinical investigational plan are listed in Table 3. Each of these is considered in the discussion below, with particular emphasis on features that may be problematic for medical devices.

4.1. Device Description

A description of the product that is the subject of the investigation is provided with sufficient information for the clinical investigators to understand the design of the device, the rationale in support of the product design (which may include reference to preclinical testing), device performance specifications, a statement of intended use, and the instructions for use.

4.2. Study Objective

Prior to initiating the design of a clinical study, it is necessary to clearly formulate the question(s) to be answered by the research effort. It should be possible to prepare a summary statement of the objective for any protocol by noting four features of the study:

Table 3
Elements of the Investigational Plan

Device description
Study objective
Study design
Study population
Treatment regimen
Control group
Endpoints evaluated
   Effectiveness
   Safety
Definition of trial success (if hypothesis-testing study)
Study procedures and duration
   Screening and assignment to treatment
   Assessments and follow-up
   Training procedures (if appropriate)
Sample size calculations
Data analysis plan
Risk analysis
Case report forms
Informed consent forms
Investigational site(s) and IRB information
Data safety monitoring board (optional)
Monitoring plan

1. The product tested;
2. The indication for use (treatment or condition to be affected);
3. The primary outcome measure; and
4. The subject characteristics (e.g., disease stage).

4.3. Study Design

The clinical study design specifies whether the study is to be prospective or retrospective, open-label (non-blinded) or controlled, and single-center or multi-center. Two groups of subjects represent the simplest controlled design, but within this general category it is possible to have many variations (e.g., crossover studies, sequential studies). More than two groups of subjects result in a multiple-arm trial, a design that may be selected to incorporate a sham as well as an active control group. It is generally best to choose the least complicated design required to successfully address the trial objective. The reader is referred for more detail on clinical trial designs to the textbook by Spilker on this topic.[9]

4.4. Study Population

In addition to articulating the clinical condition of the subjects, demographic criteria are often included to specify age, sex, and race. Economic status or educational level may be relevant, for example, in trials evaluating the labeling and instructions for use of over-the-counter in vitro diagnostics. Defining specific subject inclusion and exclusion criteria is an important means of narrowing the range of subjects studied, thereby minimizing the impact of uncontrolled variables and variability in the effectiveness and safety endpoints observed. For a pivotal clinical trial intended to support approval by the FDA, the definition of the patient population is significant because the approved PMA will generally have a label that is supportive only of the study population evaluated.

4.5. Treatment Regimen

The nature of the intervention under study encompasses a description of the device, instructions for use, and any other ancillary or related procedures or treatments. For surgical trials, it is especially critical to work with clinical investigators at all sites to develop a consensus to the greatest extent possible on surgical procedures to be used. Uniformity of procedures across sites, coupled with training of clinical investigators and their staff, are the two most critical techniques used to reduce variability and site-specific bias. For devices that represent a truly novel innovation and for which investigators are not expected to have significant first-hand experience, trials can include a prespecified run-in period to stratify sequential procedures as a function of investigator experience, and thereby evaluate the impact of the learning curve on device performance.

4.6. Control Group

The control group serves as a benchmark against which the safety and effectiveness of the device is gaged, and via this comparison, it is possible to develop an estimate of the clinical significance of the use of the device in a defined set of patients. By definition, the control group is comprised of some set of patients or subjects that are not exposed to the intervention or device.

The two broad categories of control groups are concurrent controls, and nonconcurrent controls. Concurrent controls are subjects assigned to a control exposure and observed contemporaneously with the experimental group. The control exposure may be no treatment at all, treatment with a placebo, or treatment with an alternative therapy or device. Concurrent controls are always the preferred choice, if feasible. With a concurrent control group, the subjects in the trial can be randomized between control and treatment groups, thereby eliminating selection bias and controlling for confounding variables.

Nonconcurrent controls are subjects who are not observed contemporaneously with the experimental group. The most common nonconcurrent control is the historical control group, comprised of a cohort previously observed, treated, studied, or reported on. Historical control groups must be used with caution. The quality of the historical data set is often variable and may be unreliable. The subjects in the historical cohort may not be comparable to the those in the treatment group, either in terms of demographics or disease status. There is no control for confounding variables resulting from the lack of randomization, and because the nature and success of medical treatments tend to improve over time, historical cohorts are more likely to bias the results of a trial toward a positive outcome.

The choice of the proper control group is driven by the objectives and should be supported by a sound rationale, but it is to be expected that ethical and practical constraints impact on the final choice. It is not uncommon to discover that the ideal control group from a scientific point of view is not always a feasible option for logistical or ethical reasons (e.g., a sham-operated control in a surgical trial). Well-controlled studies of some types of implants can be conducted with the patient as their own control if an alternative device or procedure is not available. Historical controls should only be used if the quality of the data set is deemed to be reliable and valid, and if the pathogenesis of the disease under study is well understood, or if the objectives of the trial are limited, such as a feasibility study.

4.7. Effectiveness and Safety Endpoints

The protocol should specify the clinical endpoint to be assessed in the evaluation of device effectiveness. The selection of the primary effectiveness endpoint is critical to the success of the trial. It should be a clinically significant outcome that can be measured using a validated method. For a hypothesis-testing study, the extent of change compared to control should be expected to be large enough to have a clinically significant impact on the patient, in terms of quality of life, disease progression, diagnosis, or mitigation. It is most common to employ a single, primary endpoint, but some trials may incorporate multiple "primary" endpoints, with the goal being to demonstrate a clinically significant response as reflected in a concordance of several measures. Chapter 10 provides an example of the latter approach.

Data on safety and potential adverse events associated with the intervention under study are collected as broadly as possible. Previous clinical experience with the device and/or data from nonclinical studies will serve to highlight specific safety issues that should be considered in a study. However, these targeted evaluations do not preclude the need to capture all data on observed adverse events, regardless of whether they are deemed to be device-related at the time.

4.8. Definition of Trial Success

If the clinical trial is a hypothesis-testing study, it is necessary to define in the protocol the definition of treatment success. This is usually expressed in terms of the magnitude of some clinical outcome deemed to be sufficient to conclude that the device is effective, in comparison to a baseline value and/or the control-group response. In the United States, demonstrations of device effectiveness do not require that the new device be superior to existing products; instead it is acceptable for product performance to be "not worse than" the standard alternative treatment(s).

4.9. Study Procedures and Duration

This section of the protocol describes the conduct and duration of the study, including screening procedures, scope of initial patient work-ups, schedules and procedures for follow-up visits, procedures for discontinuations, and so on. Also included are the procedures for assigning subjects to the intervention (randomization scheme).

Standardized training procedures for clinical investigators in the use of the device and the study procedures are important features of well-controlled studies of medical devices. Training programs enhance the rigor of a trial by minimizing bias and the impact of confounding variables on device performance that often result from variability in individual user skill and discretion. The extent of training provided to users in a clinical trial is variable, and extensive training may appropriately raise questions regarding the extent of training that a manufacturer should provide once marketing commences. In certain trials, such as for over-the-counter in vitro diagnostics, it is appropriate to provide no training in the use of the device in order to simulate as closely as possible the actual conditions of use.

4.10. Sample Size Calculations and Data Analysis Plan

For a hypothesis-testing study, the number of subjects selected for inclusion in the study should be justified by sample size calculations that specify the statistical power of the study. In order to calculate sample sizes, it is necessary to define the number of subjects specified by the underlying statistical model selected. Usually these parameters include the magnitude of the anticipated outcome resulting from the intervention under study, the variability in the measure of that outcome, the desired power of the study (usually 80%), and the statistical significance (most often 95%). The statistical approach to be used in the analysis of the data collected is articulated in the data-analysis plan. Consultation and collaboration with an expert statistician is necessary to address these aspects of the protocol.

4.11. Risk Analysis

The risk analysis provides a discussion of the potential risks and benefits to the patients accrued to the study in sufficient detail to provide adequate informed consent, and to support the conclusion that the risks to the patients are not unreasonable. The risk analysis includes a description of alternative procedures available, a consideration of potential failure modes, the steps taken in the design of the device and the trial to minimize risks to the trial subjects, and the rationale for anticipated benefit to the patient.

4.12. Informed Consent Forms

Informed consent materials for the subjects in the trial should provide complete information on the procedures and potential risks involved in the study, in a format that is easy to read and understand. The informed consent form to be signed by the subject should be explicit regarding the voluntary nature of the subject's participation, the known and potential risks and benefits of participation, and a willingness on the part of the subject to participate in all required aspects of the study. Adequate informed consent for subjects in clinical trials is required by regulatory authorities throughout the world [17-19].

4.13. Case Report Forms

Case report forms are custom-designed for each trial to record the data collected on each subject. In most cases, more than one form is required, including a subject screening form, operative information, postoperative and follow-up visit data, adverse event reports, and subject withdrawal forms.

4.14. Investigational Site(s) and Institutional Review Board (IRB) Approvals

The investigational plan specifies the investigational site(s) involved in the study, including information on the qualifications of the clinical investigators, the institutional review board (IRB) procedures, and eventually, records of IRB approvals.

4.15. Data Safety Monitoring Board

In certain circumstances, a sponsor will elect to establish a Data Safety Monitoring Board comprised of experts to evaluate on an ongoing basis the safety data accrued in a study. This is most common for clinical investigations that represent early clinical experience with an investigational device, or studies in which significant morbidity or mortality is expected. In either case, the monitoring of safety data while a study is ongoing is meant to minimize risks to the patients by being alert to unexpected adverse outcomes that, in certain circumstances, may require that a trial be terminated prematurely.

4.16. Monitoring Plan

A clinical trial monitoring plan defines the procedures to be undertaken by the sponsor, or on behalf of the sponsor, to provide quality assurance in the conduct of the study. The monitoring plan ensures that the study is conducted in accordance with the procedures specified in the protocol. The importance of a carefully designed and implemented monitoring plan cannot be overstated. This aspect of study conduct ensures that the data generated from the trial are valid, serves to identify early in the study any significant problems that may arise, and helps to ensure that the eventual audit of the clinical study by regulatory authorities is satisfactory.

5. Conclusions

Well-controlled clinical trials of medical devices have become the standard in the industry for the premarket evaluation of new products and for systematic evaluations of performance for products already on the market. The basic principles of good clinical study design developed for trials of pharmaceuticals provide the best foundation for the design of trials of medical devices. Although it is common to discover in the design of a prospective, analytical study of a medical device that there are unexpected sources of bias and confounding that require adjustments to the clinical trial strategy, rigorous studies can be developed using innovative or alternative methods. A successful clinical investigation requires careful planning to clearly delineate a testable hypothesis and to select:

1. A study design that can support the required analysis;
2. A suitable control group; and
3. Primary outcome measures that are objective, validated, and clinically significant.

It is important to remember that some questions regarding safety and effectiveness of medical devices are not readily answered in prospective, controlled clinical studies—especially for implanted devices. Most notably, it is necessary to rely on a combination of data from bench testing, animal studies, device retrieval analysis, and observational research in order to ascertain with some degree of confidence the anticipated duration of in vivo performance, failure modes, and the long-term fate of implanted materials.

References

1. U.S. Congress. *Federal Food, Drug, and Cosmetic Act of 1938.* Public Law Number 75-717, 52 Stat. 1040 (1938), 21 USC §§201.
2. Study Group on Medical Devices. 1970. *Medical Devices: A Legislative Plan.* Department of Health, Education, and Welfare, Washington, DC.
3. Food and Drug Administration, Health and Human Services. 1995. *Determination of Safety and Effectiveness.* 21 CFR §860.7(e)(1).
4. U.S. Congress. *Federal Food, Drug, and Cosmetic Act of 1938.* Public Law Number 75-717, 52 Stat. 1040 (1938), 21 USC §§505.
5. Hutt, P. B., Merrill, R. A., and Kirschenbaum, A.M. 1992. Devices are not drugs: the standard of evidence required for premarket approval under the 1976 medical device amendments. Unpublished.
6. Food and Drug Administration. 1993. *Final Report of the Committee for Clinical Review. "The Temple Report."* FDA Report. March 1993. pp. 1–45.

7. Food and Drug Administration. 1995. Notice. Statement regarding the demonstration of effectiveness of human drug products. *Fed. Regist.* 60:39,180–39,181.
8. Moss, A. J., Hamberger, S., Moore, R. M., Jeng, L. L., and Howie, L. J. 1991. Use of selected medical device implants in the United States, 1988. U.S. Department Health Human Services. *Advance Data* 191:1–24.
9. Food and Drug Administration, Health and Human Services. 1995. *General Provisions. Definitions.* 21 CFR §812.3.
10. Food and Drug Administration, Center for Devices and Radiological Health. 1995. *Clinical Trial Guidance for Non-Diagnostic Medical Devices.* Rockville, MD, pp. 1–25.
11. Food and Drug Administration, Center for Devices and Radiological Health. 1996. *Statistical Guidance for Clinical Trials of Non-Diagnostic Medical Devices.* Rockville, MD, pp. 1–32.
12. Spilker, B. and Schoenfelder, J. 1990. *Presentation of Clinical Data.* Raven, New York.
13. Food and Drug Administration, Center for Drug Evaluation and Research. 1988. *Guideline for the Format and Content of the Clinical and Statistical Sections of an Application.* Rockville, MD, pp. 1–125.
14. Minert, C. L. 1986. *Clinical Trials. Design, Conduct, and Analysis.* Oxford University Press, New York.
15. Food and Drug Administration. 1995. *International conference on harmonization. Draft guidelines on good clinical practice.* Fed. Regist. 60: 42,948–42,957.
16. CPMP. Committee for Proprietary Medicinal Products. 1994. *Biostatistical Methodology in Clinical Trials in Applications for Marketing Authorizations for Medicinal Products.* Document No. III/3630/92-EN. European Commission, Directorate-General III, Brussels.
17. Food and Drug Administration, Health and Human Services. 1995. *Protection of human subjects.* 21 CFR §50.
18. Declaration of Helsinki. World Medical Assembly, Helsinki, Finland, 1964; 1975; 1983.
19. CEN. European Committee for Standardization. 1993. *Clinical investigation of medical devices for human subjects.* Document No. EN 540: 1993 E. European Standard, Central Secretariat, Brussels.

2

Observational Research
The Nonexperimental Approach

Rosanne B. McTyre and Linda M. Pottern

1. Introduction

The randomized controlled clinical trial, or experimental approach, is considered to be the "gold standard" for the evaluation of medical device safety and effectiveness, and constitutes the basis of the approval process for many medical devices.[1] One of the most important reasons why this method is preferred is that, through the randomization process, it is more likely to lead to an unbiased interpretation of results than other types of study designs, as described in Chapter 1. It should be recognized, however, that there are situations in which the experimental approach is either not feasible because of cost or practical limitations (e.g., a rare disease), or unethical as in circumstances where no alternative therapy exists. Occasionally, nonrandomized concurrent controlled trials or even observational studies based on sound historical data may be considered valid alternatives in the evaluation of devices.[2]

Nonexperimental studies, otherwise known as observational studies, are defined as investigations that do not involve control by the researcher of the exposure experience or treatment intervention; rather, the "observed" differences in the actual experiences of the groups under consideration are recorded and evaluated.[3] There are various approaches for conducting observational studies in the evaluation of medical devices. The specific approach depends on the question or questions being addressed, a sufficient quantity of available

From: *Clinical Evaluation of Medical Devices: Principles and Case Studies*
Edited by K. B. Witkin Humana Press Inc., Totowa, NJ

data of good quality, and the resources (e.g., time and funding) available to the investigator. There are certain minimum criteria that must be met to ensure that the interpretation of results and conclusions derived from these types of studies are valid. Ultimately, if well conducted, observational studies can be used to confirm or enhance data derived from clinical trials, to assist postmarketing surveillance efforts, to suggest new indications for a device, and to uncover new research hypotheses for further testing.

This chapter will address some of the common approaches for carrying out observational studies on medical devices. It is not the objective of this chapter to provide technical advice in designing a particular investigation, but rather to provide general guidelines that may be useful in a variety of applications.

2. Observational Research Approaches

The observational study is an investigational tool that can be a useful resource to those interested in evaluating medical devices. The first and most critical step in conducting an observational study, however, is to define the questions or hypotheses to be examined. A clear definition of the premise of the study is required to determine the appropriate study design and calculate adequate sample sizes to achieve optimal statistical power in the analysis of data. Depending on the study design that is identified as most appropriate, specific inherent methodological issues must be considered in the selection of the study population and the analysis of data. This chapter will present some of the commonly used observational research approaches, useful sources of data, and a discussion of the most important issues that may threaten their validity. Three broad types of research approaches will be discussed: the case series/clinical follow-up, case-control, and cohort study designs. An example from the published literature will be presented after each study type to illustrate the proper application of each methodology.

2.1. Case Series/Clinical Follow-Up Approach

Clinicians and other practitioners are presented with a unique opportunity to identify and follow clinical outcomes of individuals whom they encounter in their daily practice. Unusual and interesting events that occur among patients are often reported in the medical literature as case reports or case series. These reports can provide important clues to the origin of diseases and/or adverse or beneficial effects of specific treatments. For some treatments, the evidence based on clinical experience can be so overwhelming that formal testing is not necessary to prove its value (e.g., the vast quantity of data on total hip arthroplasty, which has been amassed since the 1970s, as described in Chapter 4). However, for most treatments, the effects are not as dramatic and further evaluation of clinical series data is required either by including a suitable population for comparison or by using an alternative study design.

One of the most common uses of a case series in the area of medical device evaluation has been the clinical follow-up approach; the medical literature is replete with examples of clinical follow-up studies. In these studies, patients who receive a medical device or treatment are followed over time to ascertain one or more clinical or functional outcomes. Although this approach may provide useful descriptive information, it is not an adequate methodology to evaluate the relative safety and effectiveness of the device or treatment. This is because clinical follow-up studies typically lack a control group or baseline data for comparison. Clinical case series are particularly prone to bias in subject selection, which can potentially lead to false conclusions. It is thus important to properly assess and interpret data derived from clinical follow-up studies.

One of the ways in which clinical follow-up data can be optimized is by expanding the study to include an appropriate comparison population. In selecting a comparison group, particular attention must be given to minimizing potential biases (e.g., bias in the selection of subjects) by ensuring that the populations under consideration are comparable with regard to basic characteristics (e.g., age, gender, reason for treatment). The expanded clinical follow-up study can include pre vs posttreatment comparisons where individuals serve as their own control, comparison of two different treatments for the same condition using data collected in the past (historical controls), data collected at the same time as those treated (concurrent controls), or comparison of observed events, with the expected number based on regional/local medical data, such as health maintenance organizations (HMOs) or national databases (*see* Table 1), where these data exist for the specific health outcome(s) being examined. Table 1 presents selected national databases from which suitable comparison populations may be available for the calculation of expected baseline

Table 1
Selected Ongoing National Databases

Name of database	Source	Brief description	Reference
Medical Device Implant (MDI) Survey	National Center for Health Statstics	Supplement to National Health Interview Survey (NHIS) in 1988. Estimates of the total number of medical devices implanted in the United States; information on selected generic classes of devices.	Moss, A. J., Hamburger, S., Moore, R. M., Jeng, L. L., and Howie, L. J. 1991. Use of selected medical device implants in the United States, 1988. Advance Data 191, from Vital and Health Statistics of the National Center for Health Statistics, 3700 East-West Highway, Hyattsville, MD 20782
National Ambulatory Medical Care Survey (NAMCS)	National Center for Health Statistics, Ambulatory Care Statistics Branch	Data are based on information obtained from a patient encounter form for a sample of visits provided by a national probability sample of office-based physicians. A multistage design is used involving samples of primary sampling units (PSUs), physician practices within PSUs, and patient visits within physician practices. Physicians are asked to complete records for a systematic sample of office visits occurring during a randomly assigned 1-wk reporting period.	Published data: Series 13, Vital and Health Statistics and Advance Data from the Vital and Health Statistics. Unpublished data are available both in public-use data tape form and in unpublished tabulations. Public use data tapes are available from the Scientific and Technical Information Branch, Division of Data Services, National Center for Health Statistics, 3700 East-West Highway, Hyattsville, MD 20782
National Health Interview Survey (NHIS)	National Center for Health Statistics; Centers for Disease Control; Division of Health Interview Statistics	Probability sample of the civilian noninstitutionalized US population. Self-reported incidence of acute health conditions, prevalence of selected chronic conditions, limitation of activity, hospitalizations, disability days, physician visits, and other health care matters.	Published data: Current Estimates from the Health Interview Survey, an annual publication of the basic statistics derived from the NHIS. Published as Vital and Health Statistics, Series 10, by the US Government Printing Office. Unpublished data: Data available both in public-use data tape form and in unpublished tabulations. Unpublished tabulations exist for a variety of subjects relative to health status information. Division of Health Interview Statistics, NCHS, Hyattsville, MD. Public-use data tapes can be obtained through the NCHS, Division of Health Interview Statistics.

Table 1 (Continued)

Name of database	Source	Brief description	Reference
National Health and Nutritional Examination Surveys (NHANES) I and II	National Center for Health Statistics; Centers for Disease Control; Division of Health Examination Statistics	Probability sample of the civilian noninstitutionalized US population. Household interviews followed by dietary interview, physical examination, and laboratory, radiologic, and anthropometric measures in mobile examination centers. In NHANES II, laboratory testing was more extensive and standardized.	Published data: Series II of Vital and Health Statistics and Advance Data reports. Data tapes are available; catalog is available from the Scientific and Technical Information Branch, National Center for Health Statistics, Room 1-57, 3700 East-West Highway, Hyattsville, MD 20782.
NHANES III	National Center for Health Statistics; Centers for Disease Control; Division of Health Examination Statistics	Probability sample of clusters of persons representing the civilian noninstitutionalized population, with oversampling of blacks and Hispanics. Data are collected by household interview with medical examination and laboratory testing being conducted in a mobile center. Data include medical histories, physician examination, laboratory tests, and nutritional information from 24-h food recall and food-frequency interviews. Laboratory testing and physical measurements were more extensive than NHANES II.	Published data: Vital and Health Statistics and Advance Data reports. Data tapes are available; catalog is available from the Scientific and Technical Information Branch, National Center for Health Statistics, Room 1-57, 3700 East-West Highway, Hyattsville, MD 20782.
National Hospital Discharge Survey (NHDS)	National Center for Health Statistics; Centers for Disease Control; Hospital Care Statistics Branch	Stratified probability sample of hospitals and patients within hospitals. Stage one includes a 10% sample of all short-day, non-Federal hospitals, and stage two includes a sample of discharges.	Published data: Annual reports in the NCHS Vital and Health Statistics, Series 13, and Advance Data Reports. Special reports on average length of stay and specific diagnoses are published in Series 2 and 13. Unpublished data: Data tapes are available from the National Technical Information Service (NTIS), 5285 Port Royal Road, Springfield, VA 22101.

(continued)

Table 1 (Continued)

Name of database	Source	Brief description	Reference
National Mortality Statistics File	National Center for Health Statistics; Division of Vital Statistics	Mortality data include all deaths occurring annually within the United States. Reported to state vital registration offices.	Published data: National Center for Health Statistics. Vital Statistics of the United States. Volume II. Mortality. Parts A and B, and Monthly Vital Statistics Report. Annual Vital Statistics of the United States and Vital and Health Statistics. Series 20 and 21 for state data from the Mortality Surveillance Program Data Base.
NHANES I Epidemiologic Follow-up Survey	National Center for Health Statistics; National Institute on Aging; other NIH Institutes	Follow-up performed through personal interview and hospital and nursing home records and death certificates of adults ages 25–74 who were examined in NHANES I. Changes ascertained in risk factors, morbidity, functional limitations, hospitalization, and nursing home utilization.	Public use data tapes available through NTIS. A list of tapes is available through the Scientific and Technical Information Branch, National Center for Health Statistics, Room 1-57, 3700 East-West Highway, Hyattsville, MD 20782.
Medicare Provider Analyses and Review (MEDPAR)	Health Care Financing Administration (HCFA); US Department of Health and Human Services, Bureau of Data Management and Strategy	A file of annual hospital discharges containing detailed accommodation and departmental charge data, days of care, diagnostic and surgical information, and beneficiary and hospital demographics. File includes beneficiaries who are 65 yr and over, are disabled, or who have end-stage renal disease. A primary and up to four additional diagnoses are listed.	Tapes are available from the Health Care Financing Administration; Bureau of Data Management and Strategy, 7000 Security Office Park Building, Loc 1-A-11; 7000 Security Boulevard; Baltimore, MD.
Surveillance, Epidemiology, and End Results (SEER) Program: SEER Cancer Statistics Review	National Cancer Institute; Division of Cancer Prevention Control	Estimates of cancer incidence, mortality, and patient survival in a 10% sample of the US population collected from nine SEER population-based cancer registries in various areas of the country.	Published data: US Department of Health and Human Services, Public Health Service, National Institutes of Health, National Cancer Institute, Bethesda, MD. Selected data sets are available for analysis.

events for certain outcomes. In order to select the appropriate comparison group using national databases, investigators should first familiarize themselves with the limitations of the datasets and consider the applicability of these data to their own clinical populations.

2.1.1. An Example from the Literature

An example of a clinical follow-up study that used pre- and post-treatment comparisons was a multicenter case series of artificial disk implantation recently published by Griffith et al.[4] Three European surgeons conducted a retrospective medical chart review of 93 patients who had received the LINK® SB Charité Intervertebral prosthesis. The focus of the investigation was the safety and the effectiveness associated with using the Model III disks. In contrast to earlier reports on this device, this study attempted to provide a comparison (albeit retrospective) between the patients' surgery experience and their follow-up results (i.e., a pre- and postcontrolled comparison). Various parameters (i.e., study outcomes) were used to assessed efficacy: pre and postwork status, the patient's subjective pain report (right leg, left leg, and back) using an analog pain scale when available, neurologic signs and symptoms assessed from physical exam, and maintenance of mobility. The safety of the device was assessed by evaluating the complications that were encountered, both intraoperatively and postoperatively (i.e., biomechanical failure of the prosthesis, inappropriate intraoperative prosthetic sizing).

The medical outcome or clinical efficacy of the artificial disk implantation was assessed by a statistical comparison of the preoperative and postoperative data. Results showed that after an average follow up of 11.5 ± 7.3 yr, a statistically significant proportion of the patients reported to have experienced relief in back and leg pain. Statistical improvements were also noted in pain intensity and neurological weakness. Device failure, migration, or dislocation occurred in only 6 out of 93 (6.5%) patients.

Findings from this study suggest that disk implantation may be a reasonable alternative to spinal fusion for some patients, which had been considered the "gold standard" for the treatment of spine disorders at the time of the study. The authors were careful to note, however, that this study only provided a framework for future studies of artificial disk replacement, and proposed that well-controlled prospective studies comparing disk replacement surgery with fusion in carefully selected patients using clear-cut surgical indications be undertaken to evaluate this emerging technology.

2.2. Case-Control or Retrospective Approach

The case-control design can be used in a variety of applications, ranging from problem-solving and etiologic research to the evaluation of medical interventions and therapies. In the case-control approach, comparisons are made between a group of subjects who have the outcome under investigation (cases) and a group who do not have the outcome (controls). Outcomes can be defined as an adverse health effect, disease, death, or functional status. Case-control studies may be viewed as an extension of the case series but with an essential addition: a "control group." The control group is incorporated as part of the study design, which allows a comparison to be made with regard to exposure history while controlling for potential biases and confounding. Confounding may arise if the association between exposure to a factor of interest and the consequent development of an outcome is distorted by an additional variable (confounder) that is itself associated both with the factor and the outcome of interest.[3] Obtaining data on important characteristics of the study population (e.g., demographic, personal, lifestyle factors) is therefore an important part of the case-control study methodology and allows for consideration of potentially confounding factors in the analysis.

In the case-control approach, the measure used to quantify the strength of the association between a risk factor (exposure) and outcome is the odds ratio (often abbreviated OR).[5] This ratio compares the odds of developing the outcome for exposed individuals with that of unexposed individuals, and is an estimate of the relative risk (abbreviated RR, which is described in more detail in Section 2.3).

The proper selection of cases and controls is often difficult but considered essential to the validity of a case-control study. The hypotheses of the study and logistical considerations are the driving forces behind the selection of cases and controls, as well as the assessment of exposure or exposures.

2.2.1. Selection of Cases

Once the study questions are clearly defined, the criteria for case selection must be specified, including how the cases will be identified and the sources from which they will be obtained. Cases can be identified from a variety of sources, including hospitals (e.g., from medical records, discharge diagnoses, pathology and laboratory reports), HMOs, special reporting systems like cancer registries or disease surveillance systems, and physicians' records.

Introduction of bias when choosing the case population should be avoided to the extent possible. This can be achieved by incorporating into the case selection process strategies to minimize failure of eligible subjects to participate by virtue of having the outcome (disease) under study, to maximize the representativeness of the cases with respect to the population being sampled, and to ensure objective evidence and documentation of disease diagnosis. Failure to account for possible biases in the ascertainment of cases may introduce variability across studies that may lead to inconsistencies in the interpretation of findings.

In defining criteria for case selection, care must be taken to include a broad spectrum of disease severity and to avoid using subjective assessments of disease diagnosis (e.g., ill-defined signs and symptoms) by the subject or physician. Limiting the case group to only mild or severe forms of a disease can lead to misclassification of cases as noncases if early diagnosis is imprecise, or can lead to exclusion of cases who have been cured or died before symptoms were severe enough to meet the case definition.

2.2.2. Selection of Controls

Selection of appropriate controls requires careful consideration of two important issues: the source(s) from which controls will be drawn and the methods of control selection from the identified sources. The general principle is that the controls should be derived from the same "population at risk" as the cases (i.e., should represent those who might have become cases in the study) to avoid potential biases, as described in Section 2.2.1. Ideally, controls should be selected from the same source population as the cases. For example, if the cases are selected from patients who received care at a single hospital in a large city, the controls should be drawn from the population who would be likely to utilize that hospital. When selecting controls, consideration should be given to matching controls by individual or group characteristics of the cases (e.g., age, gender, race) to maximize the comparability of case-control comparisons in the analysis. Otherwise, potential confounding can be controlled in the analysis of the data.[3, 5]

Depending on the case definition, study parameters, and logistical constraints, controls can be sampled from several sources, including hospitals or other health-care facilities, the community, the general population (e.g., by random-digit dialing techniques), or national databases. Occasionally, historical controls identified from existing data sources have been used to assess the efficacy of new therapies. As noted earlier, historical controls can be compared with a recent, often consecutive series of patients who received some new treatment. This is not the preferred methodology, however, because the comparability of data for cases and controls cannot be assured as data for cases and controls were likely collected at different time periods using different methodologies.

2.2.3. Sources of Exposure Data in Case-Control Studies

One of the advantages of a case-control study is that one or more exposures or factors of interest can be evaluated with respect to the study outcome. For the purposes of this discussion, the exposures of interest are defined as implanted medical devices or related treatment(s). Such exposure data can be obtained using various approaches, including self-administered questionnaires or interviews with subjects or next of kin, records or computerized files kept on individuals (e.g., hospital, medical, registry), national databases, and occasionally from data maintained on an ongoing basis for administrative, sales and marketing, regulatory, or health reasons.

When data are collected from questionnaires, the potential for recall bias (e.g., a differential recollection of past exposures by cases and controls) must be recognized. Bias in the recollection of information from subjects can also lead to misclassification of exposure status, when responses given by cases or controls do not reflect the true situation. Thus, whenever possible, it is important to verify exposures through other data sources (e.g., medical records, biological markers) for cases and controls.

When interviews are conducted, another bias that may distort the findings from case-control studies may arise from the interviewer's awareness of the identity of cases and controls. This knowledge may influence the structure of the questions and the interviewer's manner, which in turn may affect the responses of the interviewee. One of the best strategies to overcome this potential bias, albeit not always feasible, is to conduct "blind" interviews where the interviewer is unaware of case or control status and the exposure or exposures of interest.

Using existing data for exposure assessment may overcome the issue of recall bias and may be a practical alternative to conducting interviews. The prospective nature in which these data are ascertained (i.e., collection of exposure information prior to onset of disease), and the relatively low cost involved in retrieval of such information make it worthwhile to pursue these data sources. However, such sources may not be the most appropriate venues for obtaining exposure data of interest. A number of issues should be assessed prior to the use of existing data, including the availability and completeness of records; coverage of the time period of interest; uniformity, consistency, and sufficiency of details on exposures of interest; the potential for selection bias or preferential use of specific devices or treatments; and the ability to link the exposure data to outcome events. If a decision is made to use existing resources to obtain data on exposure, it is crucial that quality-control measures be applied to ensure objective data collection. When there is a need for data abstraction, care should also be taken to standardize and pretest abstraction procedures and forms, train abstractors, and define rules for handling missing or conflicting data.

2.2.4. An Example from the Literature

The published literature on medical devices contains various examples of case-control studies designed to examine the potential relationship between breast implantation and connective tissue diseases (CTD). A recent interesting study was presented by Hochberg at the annual meeting of the American College of Rheumatology in October of 1994.[6] In this multicenter study, 869 women diagnosed with systemic sclerosis (i.e., cases) were recruited from three university-affiliated rheumatology clinics. A total of 2061 controls from the local community were selected by random-digit dialing procedures and matched to the cases on age in three strata and race. Detailed information on exposure to breast implants as well as potential confounding variables were obtained from cases through self-administered questionnaires and from controls through standardized telephone interviews.

Results showed that only 1.4% of cases had prior augmentation mammoplasty, compared to 1.1% of the women in the control group. The median time between receiving the implants and developing systemic sclerosis was 11.5 yr, with a range of 4–25 yr. The odds ratio for the association of breast implantation with systemic sclerosis, adjusted for age, race, marital status, and study site, was 1.25, with a 95% confidence interval (CI) of 0.62–2.53 (not statistically significant). The authors concluded that the data failed to show a significant causal association between breast augmentation and the development of systemic sclerosis.

2.3. Cohort Study Approach

Cohort studies represent another approach that can be used to investigate the relationship between medical devices and specific outcomes. Groups of people, or cohorts, may be selected as the basis of this type of study because they have undergone an exposure (e.g., medical device or treatment) for which an evaluation is desired, or constitute a group associated with a particular resource that may facilitate ascertainment of exposure, follow-up, or disease experience. Examples of the latter include members of HMOs, special occupational groups, such as physicians and nurses, and geographically defined populations.

A major advantage of the cohort approach is that multiple outcomes can be measured and evaluated in relation to various exposure parameters. A cohort study typically follows the group of exposed (treated) and nonexposed (untreated) individuals either forward in time (concurrent design) or retrospectively (nonconcurrent design; that is, from a specific point in time in the past to the present), to observe who develops a disease or outcome.

In the concurrent design, exposed and nonexposed groups are usually identified at the same time that the study is being initiated; these groups are then followed concurrently with the conduct of the study. Outcome data can be ascertained either directly (e.g., by periodic examinations, interviews) or indirectly (e.g., by review of medical or hospital records, disease or death registries). As in the case-control study, criteria for defining the outcome or outcomes of interest must be established prior to the onset of the study. In addition, procedures for the documentation and verification of outcome data must be formulated and incorporated into the study design. Also, information should be obtained on general characteristics of the study groups, such as age, gender, ethnicity, and other variables of interest, in order to account for the influence of any factors that may be related to the outcome or exposure in analysis.[3]

In the nonconcurrent design, the groups are reconstructed from already existing data with records providing exposure information on all subjects at the same time in the past and throughout the period of study. These groups are then "followed" to the present for ascertainment of outcomes. The availability of recorded exposure data, prior to knowledge or development of the outcome(s) of interest, allows for objective exposure classification. In using existing data to construct and follow a cohort, care must be taken to ascertain that these are of optimal quality, as has been previously described.

The typical measure of an association between an exposure and outcome or outcomes in a cohort study is the relative risk. The relative risk is the ratio of the incidence rates of a given outcome among the exposed vs the rates in the nonexposed. The larger the relative risk the stronger the association between the exposure and the outcome. This ratio can be adjusted for the effect of other important variables that may influence the outcome (i.e., confounders). Another measure of association used in a cohort analysis is the attributable risk, which is typically calculated as the incidence of the outcome in the exposed subjects minus the incidence of the outcome in those who are not exposed. Thus, it is the proportion of the outcome in the exposed individuals that can be attributed to the exposure.[7]

2.3.1. Sources of Exposure Data in Cohort Studies

A crucial aspect in the design of a cohort study involving medical devices is the ascertainment of the parameters of exposure. In constructing a cohort, it is not sufficient to simply ascertain whether a patient has received a medical device or related treatment. Other factors, such as reason for implantation, number of implants, and duration of implantation of each device (if more than one device was received); the manufacturer, model, and type of device(s); and the occurrence of adverse reactions and complications, are required for a valid characterization of exposure status and interpretation of the study findings. If subjects cannot be accurately categorized with respect to various exposure parameters, then such a study may not be feasible.

Sources of exposure information for the cohort study parallel those of the case-control study, as described earlier. In the nonconcurrent study design, exposure data relating to medical devices and interventions are most likely to be obtained from existing records from health-care facilities or physicians. For the concurrent study, possible sources of exposure data include baseline and periodic self-administered questionnaires or interview of cohort members, medical examinations and laboratory tests, and medical records. For example, the Nurses' Health Study is a large concurrent cohort study (described in more detail in Section 2.3.3.) that was designed to assess the relationship between silicone breast implants and the risk of CTD and related symptoms.[8] In this study, information was collected for a variety of medical conditions/procedures, including CTD and breast implantation, by self-administered questionnaires and "blinded" review of medical records.

2.3.2. Follow-Up Strategies

The greatest challenge in carrying out a cohort study is to achieve maximum follow-up of the study population so that completeness in the determination of outcomes is equal in all exposure categories. Procedures for collecting outcome data may vary with the timing of the study and the specific outcome(s) under consideration. In the concurrent cohort study, the subject's address and other information are ascertained at the onset of the study and procedures can be instituted to maintain contact via questionnaires or clinic visits at predetermined times over the duration of the study. This minimizes the loss of subjects resulting from change of location and assists in the rapid identification and verification of the outcomes of interest. Simultaneous follow-up of outcome events can be achieved, when appropriate, through the use of central repositories of data (e.g., HMOs, cancer registries, and national mortality databases).

The issue of ascertainment of outcomes over time is more difficult in the nonconcurrent cohort study since subject information (home address and other identifying information) are only available from records that may date back many years. Tracing activities for the whole cohort must include a variety of resources (e.g., credit bureau, employment records, motor vehicle administration records, and personal contact with relatives, coworkers, and friends) to ensure that every effort has been made to follow the cohort as completely as possible. Inadequate and incomplete follow-up can result in biased data and misinterpretation of study findings.

2.3.3. An Example from the Literature

The Nurses' Health Study is a good example of a recently published concurrent cohort that has evaluated the association between breast implants and CTD and related symptoms.[8] The cohort, assembled in 1976, consisted of 87,501 registered nurses followed concurrently over a period of 14 yr from 1976 to 1990. From this cohort, a total of 1183 women reported (on a standardized questionnaire that was administered in 1992) to have a breast implant between 1976 and 1990 and to have been free of CTD at implantation. CTD was identified and confirmed through a multistep procedure and "blinded" medical record review. Overall, 516 women were confirmed as having CTD from the results of a screening questionnaire containing 30 questions on symptoms of CTD based on defined criteria.

The incidence of diagnosed CTD in the group of nurses who had received implants ($n = 1183$) was compared to that of nurses who had not received implants ($n = 86,318$). The resulting relative risk for definite CTD was 0.6 (95% CI:0.2, 2.0) for women with any type of breast implant, and 0.3 (95% CI:0.0, 1.9) for women with silicone breast implants. No relationships were observed when specific types of CTD (e.g., rheumatoid arthritis, systemic sclerosis) were examined. The authors concluded that there was no association between breast implantation and CTD.

3. Discussion and Recommendations

The randomized controlled clinical trial is widely endorsed as the method of choice in the evaluation of the safety and effectiveness of a medical device. However, the observational study, when carefully designed and conducted, may not only be a useful adjunct to the clinical trial but may also play a significant role in the evaluation and monitoring of medical devices after they enter the market. Because, by definition, the investigator cannot control the assignment of exposure of treated and untreated groups, a myriad of factors may interfere with the valid interpretation of results in an observational study. Thus, it is imperative that these types of investigations be carefully designed and executed, and that data be analyzed and reported in a responsible manner. Inappropriate use of this study methodology may lead to unfounded and invalid conclusions with regard to the safety and effectiveness of medical devices or treatments.

The crucial first step in the design of any observational investigation is a clear definition of the question(s) to be answered and translation of these questions into testable hypotheses. *A priori* definition of the goals of the study is required for proper determination of the type(s) of studies that can be conducted, the types of data that will be required, and the necessary sample size to ensure adequate statistical power in the analysis of the resulting data. As discussed in earlier sections, a variety of observational study designs may be well suited to investigations involving medical devices; specific study designs (e.g., clinical follow-up, case-control, cohort) depend on the research question, the availability of appropriate data, and the logistical resources and constraints of the investigation.

One of the most readily available resources in the design of observational studies is routinely collected data. However, it is important to note that although existing data may be a useful resource in the design of these types of studies, these sources of data must be utilized appropriately because of their potential limitations.

Whether a study is based on existing data or whether it involves new data collection efforts, a good observational study is one that is designed to minimize bias and confounding. As previously explained, several types of biases (e.g., selection, ascertainment, recall) can be important threats to an observational study, and if present, could interfere with the conclusions derived from the data. As previous sections have explained, there are various ways to minimize bias and confounding at the outset of the study. Careful consideration of these issues should be undertaken before an investigation is initiated.

Because of the complexity of designing a good observational study, the medical device expert, surgeon, or medical doctor involved in the care of patients should not generally attempt to carry out these types of studies in isolation. Rather, a multidisciplinary approach to the design of observational studies is recommended. In addition to the medical specialist or other principal investigator, the ideal team should also include a biostatistician and an epidemiologist, who should be involved as part of the project team in the planning, execution, and analysis phases of the study. A biostatistician will be key in ensuring that the study includes enough participants to provide adequate statistical power to yield statistically meaningful results. In cases where a large enough sample cannot be achieved from one population group, the biostatistician may recommend that more than one center be used for the recruitment of study subjects or controls to achieve a sufficient sample. An epidemiologist's input is crucial in the design aspects of the study and study instruments, the identification and minimization of potential bias and confounding, and quality assurance.

No matter how well the study is designed and planned *a priori,* however, issues may arise during the course of the execution of the study that will have to be considered as data collection develops, and decisions may be made during the process of data collection that may alter the study's protocol as it was first conceived. As with experimental studies, no observational study is flawless, despite best efforts by the investigators to conceive and execute the study as carefully as possible. It is the responsibility of the project investigators, however, to consider foreseen and unforseen methodologic issues in the analysis of their data before the data are reported and during the course of the study, and to report caveats and limitations in their reports of findings. The epidemiologist and biostatistician will have a fundamental role in ensuring that proper interpretation of findings and responsible reporting of data are achieved.

4. Summary

In summary, observational studies are feasible and worthwhile when a testable research hypothesis can be specified, when good quality data are available, or when requisite data can be newly collected for testing the hypothesis of interest in an unbiased manner, when a suitable comparison population can be identified, and when enough individuals can be assembled to ensure reliable results. If properly utilized, the observational study approach can be cost- and time-effective and practical not only in enhancing, but also in extending findings on medical devices or treatments obtained from clinical trials.

References

1. Food and Drug Administration, Health and Human Services. 1994. *Determination of Safety and Effectiveness.* 21 CFR §860.7.
2. Sapirstein, W., Alpert, S., and Callahan, T. J. 1994. The role of clinical trials in the Food and Drug Administration approval process for cardiovascular devices. *Circulation* 89:1900–1902.
3. Lilienfeld, D. E. and Stolley, P. D. 1994. *Foundations of Epidemiology.* 3rd ed. Oxford University Press, Oxford.
4. Griffith, S. L., Shelokov, A. P., Buttner-Janz, K., LeMaire, J.-P., and Zeegers, W. S. 1994. A multicenter retrospective study of the clinical results of the LINK® SB Charité intervertebral prosthesis. The initial European experience. *Spine* 19:1842–1849.
5. Thompson, W. D. 1994. Statistical analysis of case-control studies. *Epidemiol. Rev.* 16:33–50.
6. Hochberg, M. C., Perlmutter, D. L., White, B., Steen, V., Medsger, T. A., Weisman, M., and Wigley, F. M. 1994. The association of augmentation mammoplasty with systemic sclerosis: results from a multicenter case-control study. Abstract presented at 58th Annual Meeting of the American College of Rheumatology, October 23–27, Minneapolis, MN.

7. MacMahon, B. and Pugh, T. F. 1970. *Epidemiology Principles and Methods*. Little, Brown, Boston.
8. Sanchez-Guerrero, J., Colditz, G. A., Karlson, E. W., Hunter, D. J., Speizer, F. E., and Liang, M. H. 1994. Silicone breast implants and the risk of connective-tissue diseases and symptoms. *N. Engl. J. Med.* 332:1666–1670.

3

Choosing and Evaluating Outcome Measures for Clinical Studies of Medical Devices

Selma A. Kunitz, Michele Gargano, and Rene Kozloff

1. Introduction

Deficiencies in recent medical device applications to the US Food and Drug Administration (FDA) have resulted in efforts to provide device manufacturers with guidelines for producing Premarket Approval Applications (PMAs) that contain scientific evidence for device safety and effectiveness.[1] Traditionally, clinical trials are recommended by the FDA for evaluation of the safety and effectiveness of diagnostic and/or therapeutic interventions. Clinical trial methods include a statement of a hypothesis, definition of a target population group, identification of patient and intervention characteristics (independent variables), and well-defined objective outcome measures (dependent variables or endpoints). Outcome measures, then, are a key component in the materials that the FDA uses to review and assess the safety and effectiveness of new devices.

Outcome measurement is a fertile area of research with a great deal of discussion regarding the domains encompassed, mechanisms for measurement, mode of administration (health care practitioner or self), and the appropriate measures themselves. Also actively debated is the issue of whether there should be a global outcome measure or

From: *Clinical Evaluation of Medical Devices: Principles and Case Studies*
Edited by K. B. Witkin Humana Press Inc., Totowa, NJ

disease and/or age-specific measures. Both arguments have merit and deserve careful consideration. In addition to the interest in outcome measures for clinical trials, researchers interested in medical effectiveness and health services research are contributing to the development of outcome measures.

Clinical outcome measures have traditionally focused on clinical events, such as mortality. However, since an increasing proportion of health-care dollars are spent on care for patients with chronic diseases, therapeutic goals have evolved to include reduced morbidity and improve patient function. Improved morbidity and function can be measured through reducing symptoms or severity of illness or limiting disease progression. However, researchers and clinicians recognize that for patients with chronic disease, biomedical measures alone do not suffice. Instead, sociomedical measures of direct importance to the patient must also be taken into consideration; such measures concern activities of daily living functions, productivity, social roles, intellectual capacity, emotional stability, and life satisfaction. These sociomedical measures are often grouped into the term "quality of life" and are used with increasing frequency as outcome measures in current clinical trials. Finally, with the national interest in containing health-care costs as well as the growing number of competing therapeutic interventions, interest in the economic impact of new medical devices is increasing. Methods for identifying and assessing economic impact are pursued in the pharmacoeconomic arena.

This chapter provides a discussion of the types of outcome measures, along with their features and limitations, selection criteria, and the relationship of those outcome measures to a device application. Outcome measures used in some recent studies are identified, along with suggestions for additional such measures in future studies of the safety and effectiveness of medical devices.

2. Outcome Measures

In recent years, there has been an increase in the range of available outcome measures. Until the 1970s, most clinical studies defined outcome as a dichotomous variable, "dead" or "alive." As clinical trials began to test interventions for chronic diseases, measures of physical function evolved and are used in many of today's drug-treatment studies. Such measures include activities of daily living[2] and disease-specific scales for cancer (Performance Status Scale, Quality of Life

Index);[3, 4] arthritis (American Rheumatism Association Classification);[5] heart disease (Specific Activity Scale);[6] depression (Beck's Depression Inventory, Zung's Depression Scale, Hamilton's Depression Scale);[7-9] and cognition (Folstein's Mini-Mental State).[10, 11]

When questioned, many patients feel that outcome measures relating to the reduction of symptoms, improvement in daily functioning, or improvement in a sense of well-being and health-related quality of life are more important than death as outcomes.[12] Thus, multidimensional measures that capture these outcomes were developed and refined in the health status assessment arena. The Sickness Impact Profile[13-15] and the Index of Well-Being[16] are sound psychometric measures. The McMaster Health Index Questionnaire[17], the Duke-University of North Carolina Health Profile,[18] and the RAND Health Insurance Experiment (HIE)[19] provide multidimensional measures of outcome. Other instruments have been developed based on a few single-item measures. For example, Spitzer et al. developed a Quality of Life index that aggregates five single-item measures of health and health-related concepts in a 1-min questionnaire.[4] However, single-item measures are often less reliable and less valid. As a compromise between long and short instruments, Stewart and her colleagues developed a general, short health survey of function and well-being that is reliable and that has evolved into the widely used SF-36.[20] All of these measures have been used in clinical trials, often under the rubric of "quality of life."

Health status assessments encompass measures that are used in clinical care and policy research as well as in clinical studies. Quality of life and health status measures are often used as interchangeable terms. One recent conceptualization of both terms established that quality of life, along with biological (including death, disease, and disability), physical (including self care, mobility, and physical activity), mental (cognitive and emotional status), and social considerations (personal and community interactions) should be considered as domains within health status assessments.[7]

3. Outcome Measurement in Medical Device Studies

Outcome measures are integral to the success of medical device studies. When well defined, they can provide sponsors and FDA reviewers with evidence for the safety and benefits to health from the use of a device.

Medical devices fall into several categories: implanted vs non-implanted devices; single-use (disposable) vs nondisposable devices; diagnostic vs therapeutic devices; energy-emitting vs nonenergy emitting devices; and life support vs nonlife support devices. Different types of medical devices may require different considerations for assessment of patient outcome.

Although development of outcome measures for clinical trials on medical devices is a relatively new endeavor, characteristics of these measures can be adapted from the drug treatment clinical trials literature. Meinert[21] identifies a set of four characteristics for primary outcome measures:

1. Outcome measures must be clinically relevant and specific (i.e., in population "W," is cardiac device "A" more effective than existing technology in reducing mortality and cardiac events post surgery?);
2. They must be easy to diagnose and observe in all subjects;
3. They must be free of measurement error and capable of unbiased assessment; and
4. They should be observed independent of treatment group and capable of being ascertained.

These characteristics clearly focus on the clinical effects of the treatment intervention on the patient. In contrast, outcome measures for medical device studies generally focus on the technological effectiveness of the product.[22] Both the impact of a device on the patient as well as its technological effectiveness may be relevant in a device clinical trial. For example, the design of a hypothetical clinical trial for a device used for dialysis will need to consider outcome measures that address both the patient quality of life (the effect on the patient's well-being), as well as the technical performance of the device (the effectiveness of removing toxins from the blood). Thus, outcome measures for a particular device should encompass the effect on an individual as well as its safety and technical performance. Additionally, the effects on the user of the device as well as the skill of the user in applying the device may affect its use and may need to be addressed.

4. Use of Medical Device Outcome Measures

A review of 52 clinical studies of medical devices that were published in 1994–1995 was conducted (*see* Table 1). This review revealed that outcome, when measured, encompassed several perspectives and

Table 1

Recent Studies—Outcome Measures Used and Potential Additional Measures[a]

Study/device, citation	Outcome measures used in a study	Potential additional measures
Pedicle screw fixation of lumbrosacral spine[29]	Postoperative CT and film radiography, physician satisfaction	
Fiberoptic bronchoscope[30]	Visualization of epiglottis without aggression or stimulation of pharyngolaryngeal structures and without modification of child's body	Quality of life
Rebreathing of CO_2 by ventilated patients with a non-breathing valve[31]	CO_2, tidal volume, and positive end-expiratory pressure levels	
Bard CT guide system[32]	Diagnostic accuracy, number of times needle repositioned	Quality of life
Insertion of second device after Rushkind occlusion system[33]	ECG/clinical improvement, no medications, and decreased cardiomegaly	Clinical status, function
Surgical limb lengthening[34]	Cosmological, functional, and psychological benefits, but no discussion of specific outcome measures	Quality of life, function
Spielberg device for ICP monitoring[35]	"Clinical outcomes"	Clinical status
Muscle-training EMG device for intractable constipation[36]	Symptomatic improvement, decrease in pelvic floor muscle activity	Clinical status, function, quality of life
Ahmed glaucoma valve implant[37]	Intraocular pressure and no additional glaucoma surgery or visually devastating complications	Clinical status, function, quality of life

(continued)

45

Table 1 (Continued)

Study/device, citation	Outcome measures used in a study	Potential additional measures
Noninvasive positive pressure ventilation[38]	Measurements of dyspnea and respiratory pressures, need for intubation	Clinical status
Home uterine activity monitoring device[39]	Cervical dilation at the time of diagnosis of preterm delivery	Clinical status, quality of life
Metered dose inhaler for asthma[40]	Asthma severity score, peak expiratory flow rate, oxygen saturation	Clinical status, function, quality of life
Baerveldt implant[41]	Postoperative visual acuity	Clinical status, function, quality of life
Airway management during pediatric myringotomy[42]	Insertion time, perioperative oxygen saturation, time from surgery completion to eye opening, response to commands and home readiness	Clinical status, function
Effects of home physiotherapy in patients with spondylitis[43]	Spinal mobility measured by fingertip-to-floor-distance	Clinical status, function, quality of life
Expansile metal stent for palliation of esophageal obstruction[44]	Dysphagia score, Karnofsky score	Function, quality of life
Spinal cord stimulation[45]	Reduction of medication, lifestyle change, improvement in activities of daily living	
Role of telemetry in guiding patient management decisions[46]	Outcomes recorded, physicians' assessment of telemetry assistance for patient management	

Table 1 (Continued)

Study/device, citation	Outcome measures used in a study	Potential additional measures
Pressurized aerosol inhaler and two breath-actuated devices[47]	Patient acceptance and inhalation technique	Function, quality of life
Prevention of pharyngospasms in tracheoesophageal speakers[48]	Swallowing function/oropharyngeoesophageal swallow efficiency	Clinical status, function, quality of life
Total hip replacement: non-cemented femoral fixation[49]	Disease-specific, global, and functional capacity outcome measures (not otherwise specified)	Quality of life
Gauze vs two polyurethane dressings for site care of pulmonary artery catheters[50]	Incidence of pulmonary artery catheter-related bloodstream infection	
Spinal cord electrical stimulation in critical limb ischemia[51]	Limb survival, patient survival, quality of life, and cost-effectiveness	

[a] Potential additional outcome measures include clinical status (parameters that describe the patient's disease-related symptoms); function (parameters that describe a person's daily life activities, often in terms of physical activities and cognitive, emotional, social, and role function); and quality of life (parameters that describe satisfaction with life, return to function, pain or discomfort, and general well-being, from the patient's perspective).

47

assessments, but the outcome measures were not clearly defined or systematically assessed. Some studies evaluated only the performance of the device, either objectively or subjectively. Physician assessment, performance ratings of the device, and assessment of how the device facilitated patient function were also used to measure outcome; these assessments were often not operationally defined or systematically measured, and were not clinically relevant or patient-centered. It is clear that the medical device studies could benefit from the more comprehensive utilization of the principles of good study design that have been designed and applied to pharmaceutical clinical trials. Methods may need to be refined or developed to address some of the unique needs of medical device studies (*see* Chapter 1).

Several studies included measuring the change in the clinical state or function of the patient after implantation of the device, and the measures often described the specific technical goal of the device, such as measuring "expiratory capacity." These outcomes demonstrate effectiveness of the device (comparison of the new device with current on-market devices, performance of the device according to specifications, and/or measurement of a clinically relevant outcome) as required for FDA approval. None of the studies reviewed utilized measures that may be more important to the sponsor and the patient, such as assessment of the patient's ability to return to activities of daily living, social, emotional function, role performance, or quality-of-life measures. Only a few studies took "patient satisfaction" into account.

In this rapidly changing and more rigorous regulatory environment, consideration should be given to expanding outcome measures for medical device studies from technical performance to clinical relevance and, in some cases, patient quality-of-life measures. For sponsors, in fact, quality-of-life measures may be most useful for device acceptance and marketing. Outcome measures that should be considered for medical device studies are:

1. Clinically relevant characteristics, such as "improved breathing;"
2. Functional performance, such as improved physical activity; and
3. Patient perceptions of quality of life, such as satisfaction with life.

Table 1 includes the studies reviewed and a brief description of the outcome measures used. Based on the outcome measures used in each study, some additional outcome measures that could be used to describe clinical relevance and quality of life are proposed:

1. Clinical status—parameters that describe the patient's disease-related symptoms;
2. Function—parameters that describe a person's daily life activities, often in terms of physical activities and cognitive, emotional, social, and role function; and
3. Quality of life—parameters that describe satisfaction with life, return to function, pain or discomfort, and general well-being from the patient's perspective.

5. Measuring Outcome

The field of outcomes research continues to evolve, and approaches to studying outcome varies among researchers and disciplines. In selecting outcomes, the following themes must be addressed: the scope of the outcome measurement, mode of assessment, domains to be assessed, economic impact, utility measures, temporal issues, and selecting appropriate outcome for measurement.

5.1. Scope of the Outcome Measurement

Outcome measures can be generic to overall health and illness, or they can be specific to a disease, condition (e.g., nausea), or body system (e.g., genitourinary system). Generic measures are those that are broadly applied across type and range of severity of disease, different medical treatments, and different patient populations. They are necessary for comparison of outcomes across different populations and interventions, and are particularly useful for cost-effectiveness studies. Disease-specific measures assess the special conditions and concerns of diagnostic groups. Specific measures may be more sensitive for the detection and quantification of small changes that are important to physicians and their patients. The choice of which approach to use is dependent on study objectives, methodological concerns, and time, resource, and financial constraints.[23] More than one type of method of outcome measure can be used in each study, and data collected by one specific variable may be used in multiple analyses. The strengths and weaknesses of the generic- and disease-specific measures are shown in Table 2.

Four different models for outcome assessment are traditionally recognized: separate generic and disease-specific measures, modified generic measures, disease-specific supplements to generic measures, and batteries in which collections of specific measures are scored

Table 2
Generic Vs Disease-Specific Outcome Measures

Type of measure	Strengths	Weaknesses
Generic	Single questionnaire	May not focus on area of interest
	Generally tested for reliability and validity in different populations	May not reflect small changes that may be important to clinicians and patients
	Broadly applicable across types and severity of diseases, across medical treatments, and across patient populations	
Disease-specific	Clinically sensitive and more responsive to the special conditions of a well-defined group	Cannot compare across condition or population

independently and reported individually. In choosing an outcome assessment approach for a particular study, the validity of content and construct, reliability, responsiveness, and generalized application of these various approaches must be considered.

5.2. Mode of Assessment

Outcome can be measured objectively or subjectively by the patient, physician, other health-care providers, or the patient's caregiver. Objective assessments of outcome include recording changes in laboratory values or in performance on written or oral tests. Progress may also be measured using subjective perceptions of changes in physical or emotional state. Again, the choice of assessment mode is dependent on the objectives of the study and the population and diseases or conditions being studied. Many researchers find it useful to combine these approaches when assessing outcome.

5.3. Outcome Domains

Outcome measures for therapeutic trials of both drugs and devices are driven by clinical relevance, objective measures, and properties that accommodate statistical analysis. Quality of life is a relatively new domain (specific area of focus) for measuring outcome that has evolved because of interest in patient satisfaction. As shown below

Table 3
Outcome Measure Domains and Instruments of Measure

Domain	Focus	Instruments of measure
Clinical status	Death, disease, impairment	Mortality, blood pressure readings, computed tomography, magnetic resonance imaging, electrocardiograms, clinical exams, and scales
Function	Self-care, mobility, physical activity, cognitive ability, emotional status, social interactions	Activities of Daily Living Scales, Neuropsychological Tests, Sickness Impact Profile, Health Status Measures
Quality of life	Satisfaction with life, sense of well being	Quality of Well-being Scale, SF-36 Health Survey, disease-specific Quality of Life Scales

and in Table 3, domains contain characteristics that are measured by specific instruments to describe patient outcome in a given study. The three major clinical trial outcome domains are as follows:

1. Clinical state and/or impairment is a clinical description that is disease-related and encompasses treatment-related symptoms. "State" or status has traditionally been described dichotomously as "dead" or "alive." More recent attempts have been made to refine this assessment. For example, patients with cardiac arrhythmias who receive pacemakers have a good outcome if their heart rhythm returns to normal.
2. Function encompasses physical activities that a person performs daily, as well as cognitive, emotional, social, and role functions. Physical function is most often measured using an Activities of Daily Living scale. Cognitive function can be measured with specific cognitive tests or a battery of tests. Emotional function can be measured by depression scales, and social function is included in health status measures, such as the SF-36 or Quality of Well Being scale. For example, if an implanted pacemaker produces a regular rhythm and the heart is working more efficiently, the patient may be able to perform better according to the Activities of Daily Living scale and return to his or her role functions.
3. Health-related quality of life is a widely discussed descriptor with an evolving definition. Health-related quality of life and health status are terms that are frequently interchanged, as are their domains. Critics of quality-of-life measures point out that they are subjective since they reflect the patient's point of view. However, disease-specific questionnaires and measures have been used in the areas of cardiac and benign

prostate hypertrophy clinical trials. Using the cardiac clinical trial above as an example, if the pacemaker restores normal heart rhythm and all body organs are more adequately perfused, the patient may be satisfied with the outcome and may have an improved quality of life.

5.4. Economic Impact

Pharmacoeconomic impact is another study-outcome domain that is attracting interest from sponsors and regulatory agencies. Several approaches are associated with the pharmacoeconomic impact of a medical diagnostic or treatment intervention. One such approach calculates the direct medical cost and indirect costs, such as disease and intervention effects, on role function and productivity. Direct medical costs include units of service required that allow for adjustments for inflation, different structures, or charges actually paid.

5.5. Utility Measures

Utility measures are derived from economic and decision theory and involve the serial measurement of a patient's quality of life throughout a study. The strength of utility measures is that they are represented by a single number, equal the net impact on quality of life, and allow for a cost/utility analysis. However, the weaknesses are that they cannot examine different aspects of quality-of-life responsiveness and that there can be difficulty in determining utility values. Two approaches to utility measurement are used:

1. The Quality of Well Being scale developed by Kaplan[24-26]: Using this scale, the patient is asked a number of questions about function. According to their responses, patients are classified into a specific category with its own utility value. The classification categories and utility values are established through previous ratings using another group; i.e., a random sample of general population.
2. The Quality-Adjusted Years Scale[27] provides a single rating for a patient that encompasses all aspects of quality of life. For this measure, at various intervals, patients are asked what trade-off that they would be willing to make of the number of years in their present health state for a shorter life-span in full health. The responses are subjected to a cost utility analysis with the product being quality-adjusted life years.

Both Quality of Well Being and Quality-Adjusted Years Scale responses are linked in order to direct illness costs to derive cost impact numbers and rates. Although economic impact is sometimes placed in the "quality of life" rubric, it combines both patient and health-care system costs, and could be argued to be a separate measure.

5.6. Temporal Issues

The frequency and duration of outcome measurements vary from study to study and are dependent on study objectives, patient availability, and resources. The type of device being used is particularly important in schedule plans. For example, there is clearly a difference in the needs for outcome determination between a dilator only used during surgery and a device, such as a joint replacement, that is left in place indefinitely. A baseline measure of the selected outcome descriptor should always be conducted. Regular intervals for assessing outcome during treatment and follow-up must be assessed.

5.7. Selecting Appropriate Outcome for Measurement

Outcome measures should be incorporated into and planned for at the start of study design. In order to successfully yield the appropriate and anticipated data, outcome measures should be defined during study protocol development, collected during the study period, and analyzed together with other clinical data.

As discussed earlier, the outcomes chosen for assessment and the methods used to measure them will be determined by study objectives, study design, and sample population. Outcome must be clearly defined and measurable. Key data items for assessment of the outcome must be identified and included in the study data-collection protocol. Variables must be selected according to the likelihood of being able to collect the information needed. For instance, disabled or debilitated patients may not be able to return weekly for in-person evaluations following the treatment period. Children may not have the attention span necessary to complete an extensive battery of tests. Clinical staff may not have the time or the interest to collect frequent repeated measures during an in-patient stay.

Recent guidelines for evaluating cardiology devices have been described in the American College of Cardiology and the American Heart Association Guidelines of Percutaneous Transluminal Coronary Angioplasty. These guidelines address the spectrum of weaknesses in device studies and offer a systematic approach to design and variable selection. This is an example of a productive collaboration among sponsors, the medical community, and the FDA.

Ultimately, data from the device study will be analyzed. The choice of a particular study endpoint is crucial to the study design since the rate of occurrence of an outcome event effects the power of a study, sample size, and study duration.[28] Current questions in the analysis of outcomes data relate to developing summary scores, the measurement modality used, and interpretation of change over time.

6. Outcome Assessment and Medical Device Approval

Well-defined outcome measures can provide sponsors and FDA reviewers with evidence that the benefits to health from the use of a device are as intended, that the benefits of use outweigh the risks, and that the device will provide clinically significant effects. The performance of the device is measured in its ability to fulfill its intended function and its reliability is reflected in the consistency of device performance. Approval of the device may be limited to those areas whose outcomes are assessed.

The outcome measures must be tailored to the specific objectives of the study and must have the clinical credibility and statistical power to support the study assertions. Careful attention must be paid in selecting variables that are measurable. Whereas performance and reliability relate specifically to technical aspects of the device, safety and effectiveness relate to the relationship of the device to the patient's clinical, functional, and quality-of-life outcomes as well as pharmacoeconomic considerations for the cost of care. Outcome measures that address safety and effectiveness of the device are likely to be pivotal for FDA approval.

References

1. Food and Drug Administration, Health and Human Services. 1995. *Determination of Safety and Effectiveness.* 21 CFR §860.7(e)(1).
2. Mahoney, F. I. and Barthel, D. W. 1965. Functional evaluation: the Barthel Index. *Md. Med. J.* 14:61–65.
3. Karnofsky, D. A., Abelmann, W. H., Craver, L. F., and Buurchenal, J. H. 1948. The use of nitrogen mustards in the palliative treatment of cancer. *Cancer* 1:634–656.
4. Spitzer, W. O., Dobson, A. J., Hall, J., Chesterman, E., Levi, J., Shepherd, R., Battista, R. N., and Catchlove, B. R. 1981. Measuring the quality of life of cancer patients: a concise QL-Index for use by physicians. *J. Chron. Dis.* 34:585–597.

5. Steinbrocker, O., Traeger, C. H., and Baterman, R. C. 1989. Therapeutic criteria in rheumatoid arthritis. *JAMA* 140:659–664.
6. Goldman, L., Hashimoto, B., Cook, E. F., and Loscalzo, A. 1981. Comparative reproducibility and validity of systems for assessing cardiovascular functional class: advantages of a new specific activity scale. *Circulation* 64:1227–1234.
7. Beck, A. T., Ward, C. H., Mendelson, M., Mock, J., and Erbaugh, J. 1961. An inventory for measuring depression. *Arch. Gen. Psych.* 4: 561–571.
8. Zung, W. W. K. 1965. A self-rating depression scale. *Arch. Gen. Psych.* 12:63–70.
9. Hamilton, M. 1967. Development of a rating scale for primary depressive illness. *Br. J. Social Cain. Psychol.* 6:278–296.
10. Folstein, M. F., Folstein, S., and McHugh, P. R. 1975. Mini-mental state: a practical method for grading the cognitive state of patients for the clinician. *J. Psych. Res.* 12:189–198.
11. Greenfield, S. and Nelson, E. C. 1992. Recent developments and future issues in the use of health status assessment measures in clinical settings. *Med. Care Supplement* 30:MS23–MS41.
12. Greenfield, S., Kaplan, S. H., Ware, J. E. Jr., Yano, E. M., and Frank, H. J. 1988. Patients' participation in medical care: effects on blood sugar control and quality of life in diabetes. *J. Gen. Int. Med.* 3:448–457.
13. Bergner, M., Bobbitt, R. A., Carter, W. B., and Gilson, B. S. 1981. The sickness impact profile: development and final revision of a health status measure. *Med. Care* 19:787–805.
14. Bergner, M. and Bobbitt, R. A. 1979. The sickness impact profile: conceptual formulation and methodology for the development of a health status measure. In *Sociomedical Health Indicators* (Elinson J. and Siegman A. E. eds.). Baywood, Farmingdale, pp. 9–31.
15. Gilson, B. S., Gilson, J. S., Bergner, M. Bobbit, R. A., Kressel, S., Pollard, W. E., and Vesselago, M. 1975. The sickness impact profile: development of an outcome measure of health care. *Am. J. Public Health* 65:1304–1310.
16. Kaplan, R. M., Bush, J. W., and Berry, C. C. 1976. Health Status: types of validity for an index of well-being. *Health Serv. Res.* 11:478–507.
17. Chambers, L. W., MacDonald, L. A., and Tugwell, P. 1982. The McMaster Health Index Questionnaire as a measure of the quality of life for patients with rheumatoid disease. *J. Rheumatol.* 9:780–784.
18. Parkerson, G. R., Jr., Broadhead, W. E., and Tse, C.-K. J. 1990. The Duke Health Profile: a 17-item measure of health and dysfunction. *Med. Care* 28:1056–1072.
19. Lohr, K. N., Brook, R. H., Kamberg, G. J., Goldberg, G. A., Leibowitz, A., Keesey, J., Reboussin, D., and Newhouse, J. P. 1986. Use of medical care in the RAND Health Insurance Experiment: diagnosis- and service-specific analyses in a randomized controlled trial. *Med. Care* 24:S1–S87.

20. Stewart, A. L., Hays, R. D., and Ware, J. E. 1988. The MOS short-form general health survey: reliability and validity in a patient population. *Med. Care* 26:724–735.
21. Meinert, C. L. 1986. *Clinical Trials: Design, Conduct, and Analysis.* Oxford University Press, New York.
22. Pepin, T. J. 1992. Design of experiments made easy. *Med. Dev. Diag. Ind.* 14:100–106.
23. Patrick, D. L. and Deyo, R. A. 1989. Generic and disease-specific measures in assessing health status and quality of life. *Med. Care* 27: S217–S232.
24. Kaplan, R. M., Bush, J. W., and Berry, C. C. 1976. Health status: types of validity and the index of well-being. *Health Services Res.* 11: 478–507.
25. Kaplan, R. M. and Bush, J. W. 1982. Health-related quality of life measurement for evaluation research and policy analysis. *Health Psychol.* 1:61–80.
26. Kaplan, R. M. 1985. Quality of life measurement. In *Measurement Strategies in Health Psychology* (Karoly P. ed.). Wiley-Interscience, New York, pp. 115–146.
27. Lohr, K. N. 1989. Advances in health status assessment: overview of the Conference. *Med. Care* 27:S1–S11.
28. Friedman, L. M., Furberg, C. D. and DeMets, D. L. (eds.) 1981. In *Fundamentals of Clinical Trials*. John Wright/PSG, Boston.
29. Kalfas, I. H., Kormos, D. W., Murphy, M. A., McKenzie, R. L., Barnett, G. H., Bell, G. R., Steiner, C. P., Trimble, M. B., and Weisenberger, J. P. 1995. Applications of frameless stereotaxy to pedicle screw fixation of the spine. *J. Neurosurg.* 83:641–647.
30. Monringal, J. P., Granry, J. C., Jeudy, C., Rod, B., and Delhumeau, A. 1994. Value of fiberoptic bronchoscope in children with epiglottitis. *Ann. Fr. Aenesth. Reanim.* 13:868–872.
31. Lofaso, F., Brochard, L., Touchard, D., Hang, T., Harf, A., and Isabey, D. 1995. Evaluation of carbon dioxide rebreathing during pressure support ventilation with airway management system (BiPAP) devices. *Chest* 108:772–778.
32. Brown, K. T., Getrajdman, G. I., and Botet, J. F. 1995. Clinical trial of the Bard CT guide system. *J. Vasc. Intervention Radiol.* 6:405–410.
33. Ledesma Velasco, M., Gomez, D. F., Solorzono Zepeda, F., Alva Espinoza. C., Montoya Guerrero, S. A., Antezana Castro, J., and Arguero Sanchez, R. 1995. Residual shunts after application of the Rushkind occlusion system in closure of persistent ductus arteriosus. *Arch. Inst. Cardiol. Mex.* 65:131–136.
34. Mastragostino, S., Boero, S., Carbone, M., and Marre Brunenghi, G. 1994. Surgical limb lengthening in patients of short stature. *Rev. Chir. Orthop. Reparatrice. Appar. Mot.* 80:634–641.
35. Schwarz, N., Matuschka, H., and Meznik, A. 1992. The Spiegelberg device for epidural registration of the ICP. *Unfallchirurg* 95:113–117.

36. Koutsomanis, D., Lennard-Jones, J. E., Roy, A. J., and Kamm, M. A. 1995. Controlled randomised trial of visual biofeedback versus muscle training without a visual display for intractable constipation. *Gut* 37: 95–99.
37. Coleman, A. L., Hill, R., Wilson, M. R., Choplin, N., Kotas-Naumann, R., Bacharach, J., and Panek, W. C. 1995. Initial clinical experience with the Ahmed glaucoma valve implant. *Am. J. Ophthalmol.* 120: 23–31.
38. Kramer, N., Meyer, T. J., Meharg, J., Cece, R. D., and Hill, N. S. 1995. Randomized, prospective trial of noninvasive positive pressure ventilation in acute respiratory failure. *Am. J. Resp. Crit. Care Med.* 151:1799–1806.
39. Wapner, R. J., Cotton, D. B., Artel, R., Librizzi, R. J., and Ross, M. G. 1995. A randomized multicenter trial assessing a home uterine activity monitoring device used in the absence of daily nursing contact. *Am. J. Obst. Gynecol.* 172:1026–1034.
40. Chou, K. J., Cunningham, S. J., and Crain, E. F. 1995. Metered-dose inhalers with spacers versus nebulizers for pediatric asthma. *Arch. Ped. Adolesc. Med.* 149:201–205.
41. Lloyd, M. A., Baerveldt, G., Fellenbaum, P. S., Sidoti, P. A., Minckler, D. S., Martone, J. F., LaBree, L., and Heuer, D. K. 1994. Intermediate-term results of a randomized clinical trial of the 350-versus the 500-mm2 Berveldt implant. *Ophthal.* 101:1456–1464.
42. Watcha, M. F., Garner, F. T., White, P. F., and Lusk, R. 1994. Laryngeal mask airway versus face mask and Guedel airway during pediatric myringotomy. *Arch. Otolaryngol. Head Neck Surg.* 120:877–880.
43. Kragg, G., Stokes, B., Groh, J., Helewa, A., and Goldsmith, C. H. 1994. The effects of comprehensive home physiotherapy and supervision on patients with ankylosing spondylitis—An 8-month followup. *J. Rheumatol.* 21:261–263.
44. Kynrim, K., Wagner, H. K., Bethge, N., Keymling, M., and Vakil, N. 1993. A controlled trial of and expansile metal stent for palliation of esophageal obstruction due to inoperable cancer. *N. Engl. Med. J.* 329:1302–1307.
45. De La Porte, C. and Van de Kelft, E. 1993. Spinal cord stimulation in failed back surgery syndrome. *Pain* 52:55–61.
46. Estrada, C. A., Rosman, H. S., Prasad, N. K., Battilana, G., Alexander, M., Held, A. C., and Young, M. J. 1995. Role of telemetry monitoring in the non-intensive care unit. *Am. J. Cardiol.* 76:960–965.
47. Nimmo, C. J., Chen, D. N., Martinusen, S. M., Ustad, T. L., and Ostrow, D. N. 1993. Assessment of patient acceptance and inhalation technique of a pressurized aerosol inhaler and two breath-actuated devices. *Ann. Pharmacotherapy* 27:922–927.
48. Pauloski, B. R., Blom, E. D., Logemann, J. A., and Hamaker, R. C. 1995. Functional outcome after surgery for prevention of pharyngospasms in tracheoesophageal speakers. *Laryngoscope* 105:1104–1110.

49. Bourne, R. B., Rorabeck, C. H., Laupacis, A., Feeny, D., Tugwell, P. S., Wong, C., and Bullas, R. 1995. Total hip replacement: the case for noncemented femoral fixation because of age. *Canad. J. Surg.* 38:S61–66.
50. Maki, D. G., Stolz, S. S., Wheeler, S., and Mermel, L. A. 1994. A prospective, randomized trial of gauze and two polyurethane dressings for site care of pulmonary artery catheters: implications for catheter management. *Crit. Care Med.* 22:1729–1737.
51. Klomp, H. M., Spincemaille, G. H., Steyerberg, E. W., Berger, M. Y., Habbema, J. D., and van Urk, H. 1995. Design issues of a randomised controlled clinical trial on spinal cord stimulation in critical limb ischaemia. *Eur. J. Vasc. Endovasc. Surg.* 10:478–485.

4

Regulatory Requirements for Clinical Trials of Medical Devices and Diagnostics

Sharon A. Segal

1. Introduction

The Food and Drug Administration (FDA) requires valid scientific evidence in order to determine whether there is reasonable assurance that a medical device is safe and effective for its intended use. This valid scientific evidence consists principally of well-controlled clinical investigations. This chapter will provide information on FDA regulatory requirements for controlled clinical investigations of medical devices and diagnostics, including who must conduct a clinical trial, the submissions required, and the procedures that must be followed.

2. Who Conducts Clinical Research?

All premarket approval (PMA) devices (i.e., Class III devices that are not substantially equivalent to a pre-1976 device) and some premarket notification (510[k]) devices (i.e., devices that can be shown to be substantially equivalent to an approved device) generally require clinical research to support the determination of safety and effectiveness (PMA) or substantial equivalence (510[k]). Clinical testing is also conducted when a supplemental application is being made to extend labeling claims and/or modify the intended use of the device. Approved products that are not intended to support a regulatory submission may also undergo clinical testing.

From: *Clinical Evaluation of Medical Devices: Principles and Case Studies*
Edited by K. B. Witkin Humana Press Inc., Totowa, NJ

There are several additional situations where clinical research of devices are conducted, but FDA regulations allow for modification of the requirements in these situations. For example, a sponsor may wish to conduct a limited clinical investigation of a device (e.g., one investigation site with a limited number of subjects) that is intended to provide data on the device's feasibility for diagnostic or therapeutic clinical use. Such limited investigations are identified as feasibility studies, Phase I studies, pilot studies, prototype studies, or introductory trials. Data from feasibility studies are not considered as pivotal evidence of safety and effectiveness. Rather, these feasibility studies form a basis to finalize and confirm the device design and determine its potential for further development, as well as to help address specific safety concerns and more clearly delineate the clinical endpoints and success/failure criteria, the intended patient population, the necessary sample size, the appropriate follow-up period, and the therapeutic effect of the device. Guidance for the conduct of feasibility studies has been published.[1, 2] In the case of new medical devices that are intended for use to treat a life-threatening or severely debilitating illness for which there is no alternate diagnosis, therapy, or prevention modality, FDA has implemented an expedited review process to ensure rapid clinical testing of these devices. Expedited review is also implemented for clinical investigations that involve no more than minimal risk or for minor changes in approved research. Guidance for expedited review investigations and minimal risk device investigations is contained in the Investigational Device Exemptions Manual.[2]

Emergency situations may arise in which an unapproved device may offer the only possible life-saving alternative, but the FDA has not approved the investigation, the intended use, or the investigating physician or institution. The FDA has allowed, on a case-by-case basis, the use of a medical device on such occasions, provided that the physician later justifies to the FDA that an emergency actually existed. The conditions for justification of emergency use and after-use procedures are specified in the FDA guidance sheets.[3]

3. Investigational Device Exemption Regulations

Approved devices are not subject to regulation by the FDA, but such studies are subject to Institutional Review Board (IRB) approval. The IRB is "...any board, committee, or other group formally designated by an institution to review, to approve the initiation of, and to conduct periodic review of, biomedical research involving human subjects. The primary purpose of such review is to assure the protection of the rights and welfare of human subjects."[4] Further discussion of the IRB's role and responsibilities in the conduct of clinical trials is provided in Section 6. Clinical trials of approved devices should be conducted in accordance with Good Clinical Practices (GCPs) (*see* Section 9.).

An investigational device is subjected to testing in a clinical trial in order to evaluate its safety and effectiveness. All clinical trials undertaken for this purpose must be conducted according to the Investigational Device Exemption (IDE) regulations.[5] Certain devices may be exempt from IDE regulations: A device, other than a transitional device, introduced into commercial distribution immediately before May 28, 1976, when used or investigated in accordance with the indications in the labeling in effect at that time; or a device, other than a transitional device, introduced into commercial distribution on or after May 28, 1976 that the FDA has determined to be substantially equivalent to a device in commercial distribution immediately before May 28, 1976, and that is used or investigated in accordance with the indications in the labeling the FDA reviewed under Subpart E of Part 807 in determining substantial equivalence.[6]

The first step in obtaining approval for the conduct of a clinical trial for an investigational device is to determine whether the investigation is "significant risk" (SR) or "nonsignificant risk" (NSR). An SR device is defined as "...an investigational device that presents a potential for serious risk to the health, safety, or welfare of a subject and is an implant; or is used in supporting or sustaining human life; or is of substantial importance in diagnosing, curing, mitigating or treating disease, or otherwise prevents impairment of human health; or otherwise presents a potential for serious risk to the health, safety, or welfare of a subject."[7] An NSR device is one that does not meet the criteria for an SR device. The FDA has issued guidance for the determination of whether a device is SR or NSR.[8] SR-device clinical investigations are subject to the full requirements of the IDE regulation, whereas NSR-device investigations must comply with an abbreviated version of these regulations. The major differences between the requirements for SR- and NSR-device studies are the approval procedures and record-keeping and reporting requirements (*see* Sections 7. and 8.).

The initial determination regarding whether a device study poses a significant or nonsignificant risk is made by the sponsor, and agreed or disagreed on by the IRB at the investigational institution. The risk determination should be based on the use of the device in the proposed clinical investigation, not on the device alone. The nature of the harm that may result from use of the device must be considered when making a significant risk determination. If the potential harm to subjects could be life-threatening, could result in permanent impairment of a body function or permanent damage to body structure, or could necessitate medical or surgical intervention to preclude these occurrences, the device should be considered SR. Furthermore, if the study protocol necessitates performing a surgical procedure, the potential harm resulting from that procedure must also be considered when making the risk determination.

Even if FDA guidelines list a particular device as NSR, it may still be prudent to seek IDE approval from the FDA. Although IDE approval offers no guarantees, it provides some measure of assurance that the clinical data generated under an IDE in support of the safety and effectiveness of a device will satisfy all of the requirements for market approval. Obtaining IDE approval thus protects the device manufacturer from reliance solely on the IRB for a determination that the clinical device trial is conducted in accordance with FDA regulations and ensures FDA participation in the design of the trial. Another advantage to conducting a clinical trial under an approved IDE is that as of November 1, 1995, the Health Care Financing Administration (HCFA) approved the Medicare reimbursement of most unapproved devices used in IDE trials.

4. FDA IDE Requirements

An IDE application must be submitted to the FDA and approved prior to commencing a clinical trial on a significant risk device.

4.1. IDE Application Requirements

A clear, concise, and appropriately detailed investigational plan is essential for a successful IDE application. Although there is no set format for an IDE application, the regulations indicate the minimum information required.[9] Table 1 provides an overview of these infor-

Table 1
Information Required in IDE Applications

Information required	Examples
Identification of study personnel	Names and addresses of sponsor, investigators, IRB chairpersons, investigational institutions, and study monitors.
Background information	Complete report of prior investigations, to include: 1. Bibliography and copies of all relevant publications on clinical, animal, and laboratory testing of the device. 2. Summary of all other relevant unpublished information. 3. Statement that all studies were conducted in compliance with Good Laboratory Practices (GLP) or justification for noncompliance.
Description of proposed study	Investigational plan: 1. Name, description, and intended use of the device. 2. Objectives and duration of the investigation. 3. Protocol. 4. Risk analysis: description of risks and how they will be minimized; justification for the investigation. 5. Description of patient population (number, age, sex, condition). 6. Monitoring procedures.
Manufacturing information	Description of methods, facilities, and controls used for the manufacture, processing, packing, storage, and installation of the device.
Agreements and certifications	Examples of agreements to be signed by investigators and informed consent forms to be signed by subjects (with accompanying informational material); certification that all investigators have signed agreements and of any action taken by IRBs.
Additional device information	1. Amount, if any, charged for the device with an explanation of why the sale does not constitute commercialization. 2. Copies of device labeling.
Exclusions	Claims for categorical exclusion or an exclusion from an environmental assessment.

mation requirements. In addition, several guidance documents are available for preparing IDE applications for specific types of devices.

Sponsors are encouraged to initiate communications with the FDA division that will review their IDE early in the process of its development in order to obtain preliminary FDA review and comment. These communications may be either pre-IDE meetings and/or pre-IDE submissions. Early interactions with FDA staff provide the sponsors with advice and guidance that can be used in the development of their

IDE applications and serve to increase the sponsor's understanding of various FDA requirements, regulations, and guidance documents, thereby facilitating the submission of more complete and approvable original applications. Examples of information that may be included in pre-IDE submissions include draft clinical protocols, proposals for preclinical testing, preclinical test results, or any other information for which the sponsor desires feedback. For further reference, the FDA has issued a guidance memorandum that discusses the procedures for pre-IDE meetings and submissions.[10]

The FDA considers the information contained in an IDE application, a pre-IDE submission, and discussions from pre-IDE meetings confidential unless it is determined that such information has previously been made publicly available. Information contained in an IDE application may also be disclosed when the FDA has approved the PMA application for the device or when the device has a Product Development Protocol notice of completion in effect. Public disclosure of the safety and effectiveness information from an IDE is allowable for providing the basis for approving, disapproving, or withdrawing approval of an IDE for a banned device, or for public consideration of a specific pending issue.

4.2. FDA IDE Approval Procedures

The FDA will notify the applicant regarding the disposition of the IDE application within 30 d of receipt of the application. FDA may approve, approve with modification, or disapprove an IDE application. Grounds for disapproval or withdrawal of approval include failure to comply with any applicable requirement of the IDE regulation, other applicable regulations, statutes, or any condition of approval imposed by an IRB or the FDA; if the application or report contains untrue statements or omits required material or information; the sponsor fails to respond to a request for additional information within the time prescribed by the FDA; there is reason to believe that the risks to the research subjects are not outweighed by the anticipated benefits or the importance of knowledge to be gained; the informed consent is inadequate; the investigation is scientifically unsound; the device as used is ineffective; or if it is unreasonable to begin or continue the investigation because of the way the device is used or the inadequacy of various components of the IDE application.[11] If an IDE application is disapproved, the applicant has the right to request a hearing.

4.3. IDE Amendments and Supplements

There is no provision for amendments to an IDE application in the IDE regulations; however, FDA considers all submissions related to an original IDE that has been previously disapproved as an IDE amendment. Submissions related to an IDE that has been approved are considered supplements. Sponsors are required to submit supplements to their IDE applications if there is a change in the investigational plan that may affect the scientific soundness of the investigation or the rights, safety, or welfare of the subjects. IRB and FDA approval of a supplement submitted for this purpose is required prior to initiating such changes unless the changes are required to protect the life or physical well-being of a subject in an emergency. In the latter case, FDA must be notified within five working days of the change.[12]

An IDE supplement must also be submitted to add new institutions or facilities to the investigation. An IDE supplement to add new institutions or facilities must include certification of IRB approval, information updating the IDE application if the investigation has changed, and a description of any modifications required by the IRB as conditions of approval. IRB approval may take place concurrently with submission of the IDE supplement to FDA. However, unless a special exemption is granted by the FDA, the investigation may not commence at the new institution until IRB approval is obtained, the FDA receives certification of this approval, and the FDA approves the supplemental application.[13]

5. IDE Requirements for Approval of an NSR Clinical Trial

If a sponsor believes that the investigation he or she wishes to conduct presents a nonsignificant risk, then the proposed study must be presented with the following information to the IRB for review: an investigational plan (to include the purpose of the study, a written protocol, a risk analysis, a description of patient selection, and a description of the device, monitoring procedures, labeling, and consent materials), a report of prior investigations, and a statement describing why the investigation does not present a significant risk.

The investigation is considered to have an approved IDE under the abbreviated requirements of the IDE regulations if it is approved as a NSR by the IRB of the institution, and the sponsor may begin the clinical investigation immediately without submission of an IDE application to the FDA. The FDA is therefore not involved in the approval process of an NSR investigation. However, if the IRB determines that the investigation presents a significant risk, then the sponsor must submit a full IDE application to the FDA. A more detailed discussion of the IRB's role and responsibilities in the approval and monitoring of a medical device clinical investigation is included in the next section.

6. IRB Review and Approval

The IRB has the authority to review and approve, require modification, disapprove, or discontinue a clinical investigation. As noted above, the IRB's primary role is to assure that the rights and welfare of the human subjects participating in the investigation are protected by reviewing the study protocol and related materials (e.g., informed consent documents and investigator brochure).

The IRB considers several factors related to the protection and well-being of subjects when deciding whether to approve clinical research.[14] For example, the IRB must decide that the risks to subjects in the trial are minimized by using procedures consistent with sound research design and that do not unnecessarily expose subjects to risk, and whenever appropriate, by using procedures already being performed on the subjects for diagnostic or treatment purposes. Furthermore, the IRB must determine that risks to subjects in the trial are reasonable in relation to anticipated benefits (if any) and the importance of the knowledge that may be expected to result. Where appropriate, the research plan must make adequate provision for monitoring the data collected to ensure the safety of subjects and that appropriate additional safeguards have been included in the study to protect the rights and welfare of subjects who are members of a vulnerable group. The IRB must be satisfied that the selection of subjects is equitable, that there are adequate provisions to protect the privacy of subjects and to maintain the confidentiality of data, and that informed consent will be sought from each prospective subject or the subject's legally authorized representative and will be documented in accordance with, and to the extent required by, the Agency's informed consent regulations.[13, 15]

In addition, the IRB must continually monitor the investigation at intervals appropriate to the degree of risk, but not less than once per year.[16] They must determine which studies need verification from sources other than the investigator and, that no material changes in the research have occurred since the previous IRB review, and ensure that changes in approved research are promptly reported to and approved by the IRB. The IRB may suspend or terminate approved research that is not being conducted in accordance with the IRB's requirements.

These represent minimum requirements, and the IRB must establish its own specific procedures for monitoring continuing research within the framework of the regulations. Therefore, IRBs can impose greater and more detailed standards of protection for human subjects than those specified by the regulations.

7. Responsibilities of Sponsors and Investigators

In addition to the responsibilities of the IRB discussed above, sponsors and investigators have clear responsibilities to ensure the safe and ethical conduct of both SR and NSR clinical investigations. The responsibilities of each of these involved parties are summarized in Table 2.

In addition to the responsibilities presented in Table 2, the sponsor is also responsible for terminating all investigations or parts of investigations as soon as possible if it is determined that an unanticipated adverse device effect presents an unreasonable risk to subjects. Terminations shall not occur later than five working days after the sponsor makes this determination and not later than 15 working days after the sponsor first received notice of the effect.

8. Records and Reporting Requirements

Adequate, appropriate, and timely record keeping and reporting are integral to carrying out the responsibilities of the sponsor, the clinical investigator, and the IRB. The specific documentation required of the sponsor and the investigator(s) involved in a clinical investigation is summarized in Tables 3 and 4. As shown in Tables 3 and 4, record-keeping and reporting requirements for SR investigations are more rigorous than those specified in the regulations for

Table 2
Responsibilities of the Sponsors and Investigators

Facet of clinical investigation	Sponsor	Investigator
Investigators	Selection (obtain CV and documentation of experience); training.	N/A
Subjects	N/A	Protect the rights, safety, and welfare of subjects.
Investigational device control	Ship only to qualified investigators involved in study.	Supervise all subjects and persons involved in the investigation; return or dispose of any remaining devices.
Monitoring	Select monitors; ensure proper monitoring procedures.	N/A
IRB	Ensure review and approval.	N/A
Agreements and certification	Signed agreements from participating investigators that include assurance that: 1. Investigator will conduct study according to investigational plan, FDA regulations, FDA and IRB conditions of approval. 2. Investigator will supervise all testing. 3. Investigator will obtain informed consent.	1. Conduct study according to investigational plan, FDA regulations, FDA and IRB conditions of approval. 2. Supervise all testing. 3. Obtain informed consent from subjects.
IDE application	Submit IDE application.	N/A
Notification of IRB and FDA	Provide any significant new information (e.g., unanticipated adverse events) to IRB and FDA.	N/A

NSR investigations. A suggested format for an IDE progress report is presented in Appendix A.

Investigators are required to maintain these records for a period of 2 yr after the date that the investigation is completed or terminated, or the records are no longer required to support a PMA or a product-development protocol. If an investigator or sponsor transfers custody

Table 3
Responsibilities for Maintaining Records[a]

Records	Maintained by	
	Investigator	Sponsor
Significant risk device		
All correspondence pertaining to the investigation, including required reports	✓	✓
Records of shipment, receipt, disposition of the device	✓	✓
Device administration and use	✓	—
Subject case histories and exposure to the device	✓	—
Informed consent forms	✓	—
Protocols and reasons for deviations from protocol	✓	—
Adverse device effects and complaints	—	✓
Signed Investigator Agreements	—	✓
Any records the FDA requires to be maintained by regulation or by specific requirement for a particular investigation	✓	✓
Nonsignificant Risk Device		
Name and intended use of device	—	✓
Objectives of the investigation	—	✓
Brief explanation of why device does not involve significant risk	—	✓
Name and addresses of investigator(s) and IRBs	—	✓
Extent to which GMPs will be followed	—	✓
Informed consent forms	✓	—
Adverse device effects and complaints	—	✓

[a]Adapted from: Food and Drug Administration, Public Health Service. 1996. *Investigational Device Exemptions Manual.* HHS Publication FDA 96-4159.

of the records to another person, the FDA must be notified within 10 working days after the transfer occurs. The FDA has the authority to inspect facilities where investigations are being held, manufactured, packed, installed, used, or implanted, or where records of use are kept. Therefore, sponsors, IRBs, and investigators are required to permit authorized FDA personnel reasonable access at reasonable times to inspect and copy all records of an investigation, including records that identify subjects.

IRB records must also be maintained. Records to be stored by IRBs overseeing clinical investigations at their institutions are summarized below:

1. Membership/employment/conflicts of interest.
2. Research proposals reviewed.
3. Sample consent documents.

Table 4

Responsibilities for Preparing and Submitting Reports[a]

Type of report	When report should be submitted	Report prepared by	
		Investigators, for	Sponsors, for
Significant risk devices			
Unanticipated adverse effect evaluation	As soon as possible but no later than 10 working days	Sponsors and IRBs	FDA, investigators, and IRBs
Withdrawal of IRB approval	Within five working days	Sponsors	FDA, investigators, and IRBs
Withdrawal of FDA approval	Within five working days	N/A	FDA and investigators
Progress report	At regular intervals but no less than one per year	Sponsors, monitors and IRBs	FDA and IRBs
Deviations from investigational plan	As soon as possible but no later than 5 working days	Sponsors and IRBs	FDA
Use of device without informed consent	Within five working days after the use occurs	IRBs	FDA
Final report	Within 3 mo (investigator) or 30 working days (sponsor) after study termination or completion	Sponsors and IRBs	FDA, investigators, and IRBs
Withdrawal of FDA approval	—	N/A	IRB and investigators
Current investigator list	Every 6 mo	N/A	FDA
Recall and device disposition	Within 30 working days after receipt of a request to return, repair, or dispose of an investigational device	N/A	FDA and IRBs
Records maintenance transfer	—	FDA	FDA
Significant risk determinations	Within five working days of IRB determination	N/A	FDA

Table 4 (Continued)

Type of report	When report should be submitted	Report prepared by	
		Investigators, for	Sponsors, for
Nonsignificant risk devices			
Unanticipated adverse effect evaluation	As soon as possible but no later than 10 working days	Sponsors and IRBs	FDA, investigators, and IRBs
Withdrawal of IRB approval	Within five working days	Sponsors	FDA, investigators and IRBs
Withdrawal of FDA approval	Within five working days	N/A	FDA and investigators
Progress report	At regular intervals but no less than one per year	N/A	IRBs and investigators
Final report	Within 6 mo after study termination or completion	N/A	FDA and IRBs
Inability to obtain informed consent	Within five working days after the use occurs	Sponsors and IRBs	FDA
Recall and device disposition	Within 30 working days after receipt of a request to return, repair, or dispose of an investigational device	N/A	FDA and IRBs
Significant risk determinations	—	N/A	FDA

[a] Adapted from: Food and Drug Administration, Public Health Service. 1996. *Investigational Device Exemptions Manual.* HHS Publication FDA 96-4159.

4. Investigator progress reports.
5. Reports of subject injuries.
6. Meeting minutes (to include attendance, actions taken, votes, basis for changes or disapproval, summary of controverted issues).
7. Records of continuing reviews.
8. Correspondence.
9. Membership.
10. Written procedures.
11. Statements of significant new findings.
12. Record retention.

9. Good Clinical Practices

The term Good Clinical Practices (GCPs) refers to the federal regulations and industry-accepted standards for the conduct of clinical studies, including study design, record-keeping and reporting requirements, informed consent of subjects, collection of scientific data, and submission of information needed by regulatory bodies to evaluate the safety and effectiveness of a drug, biologic, or medical device prior to its introduction to the market. In an effort to standardize GCPs internationally, a "Guideline on Good Clinical Practice" was recently developed by the International Conference on Harmonization of Technical Requirements for Registration of Pharmaceuticals for Human Use.[17] The purpose of this guideline is "...to provide a unified standard for the European Union (EU), Japan, and the United States, as well as...Australia, Canada, the Nordic countries, and the World Health Organization (WHO)...for designing, conducting, recording, and reporting trials that involve the participation of human subjects."[17] The International Conference on Harmonization guideline sets forth 13 principles for GCPs that are listed in Table 5.

10. Clinical Studies Conducted Outside the United States

Many clinical studies of devices and diagnostics are conducted outside the United States. Exportation of investigational devices is subject to provisions delineated in Section 801(3) of the Food, Drug, and Cosmetic Act; therefore, it is necessary to obtain FDA approval prior to export. Currently, manufacturers seeking to export unapproved medical devices must submit a request to the FDA for a determination that the export of the product does not adversely affect public health and that the importing country has approved the device.

Table 5
International Conference
on Harmonization Principles of Good Clinical Practice (GCP)

1. Clinical trials should be conducted in accordance with the ethical principles that have their origin in the Declaration of Helsinki, and that are consistent with GCP and the applicable regulatory requirement(s).
2. Before a trial is initiated, foreseeable risks and inconveniences should be weighed against the anticipated benefit for the individual trial, the subject, and society. A trial should be initiated and continued only if the anticipated benefits justify the risks.
3. The rights, safety, and well-being of the trial subjects are the most important considerations and should prevail over interests of science and society.
4. The available nonclinical and clinical information on an investigational product should be adequate to support the proposed clinical trial.
5. Clinical trials should be scientifically sound and described in a clear, detailed protocol.
6. A trial should be conducted in compliance with the protocol and amendment(s) that have received prior IRB/IEC approval/favorable opinion.
7. The medical care given to and medical decisions made for subjects should always be the responsibility of a qualified physician or, when appropriate, of a qualified dentist.
8. Each individual involved in conducting a trial should be qualified by education, training, and experience to perform his or her respective task(s).
9. Freely given informed consent should be obtained from every subject prior to clinical trial participation.
10. All clinical trial information should be recorded, handled, and stored in a way that allows its accurate reporting, interpretation, and verification.
11. The confidentiality of records that could identify subjects should be protected, respecting the privacy and confidentiality rules in accordance with the applicable regulatory requirement(s).
12. Investigational products should be manufactured, handled, and stored in accordance with applicable GMP standards. They should be used in accordance with the approved protocol and amendment(s).
13. Systems with procedures that assure the quality of every aspect of the trial should be implemented.

The results of clinical trials conducted outside the United States can be used to support, and in some cases establish, the safety and effectiveness of a medical device. FDA regulations are silent regarding whether IDE regulations apply to investigational device clinical investigations conducted outside the United States, but the regulations do state that demonstration of safety and effectiveness of a device in support of a PMA may be based solely on foreign clinical data if it is carefully designed and conducted. The sponsor is required to demonstrate that the data provided are applicable to the US population and US medical practice, and that they have been collected in accordance with GCPs.[18] The sponsor should expect that the FDA will conduct on-site inspections and monitor audits of foreign studies provided as pivotal support for a US registration. Depending on the nature and quality of existing foreign clinical studies, it is possible that a sponsor may submit an IDE for a modified (e.g., single-center) US clinical trial designed to answer specific questions regarding the safety and effectiveness of a device that were not adequately addressed by the existing foreign clinical studies.

11. Diagnostics

The term, in vitro diagnostic device (IVDD) encompasses a wide array of different products that are intended for use in the collection, preparation, and examination of specimens taken from the human body.[19] IVDDs are exempt from the IDE requirements if they are noninvasive, do not require an invasive sampling procedure that presents a significant risk, do not by design or intention introduce energy into a human subject, and are not used as a diagnostic procedure without confirmation of the diagnosis by another medically established product or procedure.[20] However, IVDDs meeting these exemption criteria must still satisfy the requirements for appropriate labeling of "...investigational or research use only diagnostics."[21] For example, an IVDD intended for research use only must be labeled "[F]or Research Use Only. Not for use in diagnostic procedures."[22] Alternatively, an IVDD that is intended only for investigational use must be labeled "[F]or Investigational Use Only. The performance characteristics of this product have not been established."[22]

Because of extensive misuse and abuse of the exemptions noted above, there is considerable commercialization of unapproved IVDDs, where IVDDs labeled for investigational and research use only are being promoted, distributed, and used for "live" diagnostic purposes. The FDA is concerned that such commercialization of unapproved IVDDs may result in the widespread use of test results as the basis for patient diagnosis and clinical evaluation, when devices have not had performance characteristics properly established, leading to potentially serious adverse health consequences for patients. As a result of these concerns, the FDA published a "Draft Compliance Policy Guide (CPG) on Commercialization of Unapproved in vitro Diagnostic Devices Labeled for Research and Investigation," which attempts to set controls on the distribution and use of unapproved IVDDs.[23]

The demonstration of safety and effectiveness of IVDDs focuses on well-designed multilaboratory experiments for measuring essential analytical and diagnostic (clinical) performance characteristics, such as precision, bias, interference, analytical range, linearity, carry-over, minimum detection limit, reference interval, and other characteristics that may be specific to certain types of IVDDs. If the sponsor wishes to claim that the device can be used to diagnose certain diseases or conditions, then its diagnostic sensitivity and specificity, as well as its predictive value, must be determined. To assist IVDD manufacturers in the development and conduct of clinical trials, the FDA and the Office of Device Evaluation Division of Clinical Laboratory Devices have issued a general guidance document.[24] This document, which has not yet been finalized, contains information on study protocols, sampling methods, study site requirements, product inserts, and the responsibilities of principal investigators of IVDD clinical trials.

12. European Standards for Device Clinical Trials

"EN-540: Clinical Investigations of Medical Devices for Human Subjects" sets forth requirements for the conduct of clinical trials on medical devices that will be marketed in all European countries that are members of the European Committee for Standardization (CEN) (i.e., all EU nations as well as Austria, Finland, Iceland, Norway, Sweden, and Switzerland).[25] The purpose of EN-540 is to "...protect subjects and ensure the scientific conduct of the clinical investigation." The contents of EN-540 are similar to that of the FDA's regulations for informed consent and the IDE regulations, and it includes provisions for investigational plans, study monitoring, record keeping, and informed consent. Like the FDA's IDE regulations, EN-540 stipulates that all clinical investigations must be conducted in accordance with the Declaration of Helsinki, which provides ethical conduct requirements.

There are a number of differences between EN-540 and the FDA IDE regulations. One such difference is that the FDA IDE regulations do not apply to devices approved under the Federal Food, Drug, and Cosmetic Act, whereas the CEN standard does not exempt manufacturers of approved devices from its provisions. Also, IDE regulations permit the interstate distribution of unapproved devices to qualified investigators, whereas European investigational devices are subject to all of the same requirements of approved devices, unless they are exempted on a case-by-case basis. Finally, EN-540 does not apply to in vitro diagnostic devices, but only applies to medical devices for which clinical performance must be assessed prior to being placed on the market.

Appendix A: Suggested Format for an IDE Progress Report*

1. The basics
   - IDE number.
   - Device name and indication(s) for use.
   - Sponsor's name, address, fax, and telephone number.
   - Contact person.
2. Study progress (Data from the beginning of the study should be reported, unless otherwise indicated)
   - Brief summary of study progress in relation to the investigational plan.
   - Number of investigators/investigational sites (attach list of investigators).
   - Number of subjects enrolled (by indication or model).
   - Number of devices shipped.
   - Brief summary of results.
   - Summary of anticipated and unanticipated adverse effects.
   - Description of any deviations from the investigational plan by investigators (since last progress report).
3. Risk analysis
   - Summary of any new adverse information (since the last progress report) that may affect the risk analysis; this includes preclinical data, animal studies, foreign data, clinical studies, and so forth).
   - Reprints of any articles published using data collected from this study.
   - New risk analysis, if necessary, based on new information and on study progress.

* Adapted from: Food and Drug Administration, Public Health Service. 1996. *Investigational Device Exemptions Manual.* HHS Publication FDA 96-4159.

4. Other changes
   - Summary of any changes in manufacturing practices and quality control (including changes not reported in a supplemental application).
   - Summary of all changes in investigational plan not required to be submitted in a supplemental application.
5. Future plans
   - Progress toward product approval, with projected date of PMA or 510(k) submission.
   - Any plans to change investigation, e.g., to expand study size or indications, to discontinue portions of the investigation, or to change manufacturing practices (**Note:** Actual proposals for changes should be made in a separate supplemental application.).

References

1. Food and Drug Administration, Public Health Service. 1996. *Investigational Device Exemptions Manual.* HHS Publication FDA 96-4159.
2. Food and Drug Administration, Office of Device Evaluation. 1989. *Guidance on the Review of Investigational Device Exemptions (IDE) Applications for Feasibility Studies.* ODE Guidance Memorandum 89-1. Rockville, MD.
3. Food and Drug Administration. 1995. *Information Sheets for IRBs and Clinical Investigators.* Rockville, MD.
4. Food and Drug Administration, Health and Human Services. 1995. *Institutional Review Boards* Definitions. 21 CFR §56.102(g).
5. Food and Drug Administration, Health and Human Services. 1995. *Investigational Device Exemptions.* 21 CFR §812.
6. Food and Drug Administration, Health and Human Services. 1995. *Investigational Device Exemptions.* Applicability 21 CFR §812.2(e).
7. Food and Drug Administration, Health and Human Services. 1995. *Investigational Device Exemptions.* Definitions. 21 CFR §812.3(m).
8. Food and Drug Administration, Center for Devices and Radiological Health. 1994. *Guidance on Significant and Nonsignificant Risk Studies.* Rockville, MD.
9. Food and Drug Administration, Health and Human Services. 1995. *Application and Administrative Action.* Application. 21 CFR §812.20(b).
10. Food and Drug Administration, Office of Device Evaluation. 1995. *Goals and Initiatives for the IDE Program.* ODE Guidance Memorandum 95-1. Rockville, MD.
11. Food and Drug Administration, Health and Human Services. 1995. *Application and Administrative Action.* FDA action on applications. 21 CFR §812.30.

12. Food and Drug Administration, Health and Human Services. 1995. *Application and Administrative Action*. Supplemental applications. 21 CFR §812.35.
13. Food and Drug Administration, Health and Human Services. 1995. *Protection of Human Subjects*. 21 CFR §50.
14. Food and Drug Administration, Health and Human Services. 1995. *Institutional Review Boards*. 21 CFR §56.
15. Food and Drug Administration, Health and Human Services. 1995. *Institutional Review Boards*. Criteria for IRB approval of research. 21 CFR §56.111.
16. Food and Drug Administration, Health and Human Services. 1995. *Institutional Review Boards*. 21 CFR §56.
17. Food and Drug Administration. 1997. International Conference on Harmonization; Good Clinical Practice; Consolidated Guideline; Notice of Availability. *Fed. Regist.* 62:25,692-25,709.
18. Food and Drug Administration, Health and Human Services. 1995. *Premarket Approval of Medical Devices. Research conducted outside the U.S.* 21 CFR §814.15.
19. Food and Drug Administration, Health and Human Services. 1995. *In Vitro Diagnostic Products for Human Use*. Definitions. 21 CFR §809.3.
20. Food and Drug Administration, Health and Human Services. 1995. *Investigational Device Exemptions*. Applicability. 21 CFR §812.2(c)(3).
21. Food and Drug Administration, Health and Human Services. 1995. *Investigational Device Exemptions*. Waivers. 21 CFR §812.10(c).
22. Food and Drug Administration, Health and Human Services. 1995. *In Vitro Diagnostic Products for Human Use. Labeling for in vitro diagnostic products*. 21 CFR §809.10(c)(2).
23. Food and Drug Administration. Office of Device Evaluation. 1992. *Draft Compliance Policy Guide on Commercialization of Unapproved In Vitro Diagnostic Devices Labeled for Research and Investigation* (Draft CPG). Rockville, MD.
24. Food and Drug Administration. Office of Device Evaluation. 1994. *Points to Consider for Collection of Data in Support of In Vitro Device Submissions for 510(k) Clearance*. Rockville, MD.
25. CEN. European Committee for Standardization. 1993. *Clinical Investigation of Medical Devices for Human Subjects*. Document No. EN 540: 1993 E. Brussels.

Part II

Introduction to Case Studies

Karen Becker Witkin

The case studies that form Part II were selected to provide the reader with an illustrative collection of high quality clinical research on medical devices. Each chapter describes a case study, or a related series of case studies, that effectively incorporated the principles of good clinical study design within a clearly articulated research objective and the current state of the science. In addition to meeting a rigorous scientific standard as part of the selection criteria, the case studies were also chosen to reflect significant contributions to the field on a variety of medical devices, as well as successful approaches to solving a variety of clinical research problems. The chapters that follow include investigations on orthopedic, cardiovascular, soft tissue, and in vitro diagnostic products within the context of pilot studies, pivotal trials of safety and effectiveness, postmarket surveillance investigations of long-term or rare complications, and failure analysis studies.

For pivotal trials of safety and effectiveness, the reader is referred to Chapter 5, Chapter 6, Chapter 9, and Chapter 10. These chapters illustrate that a range of clinical study designs can be relied on to generate valid data for use in formulating a sound risk/benefit analysis of device performance within the context of existing clinical practice.

From: *Clinical Evaluation of Medical Devices: Principles and Case Studies*
Edited by K. B. Witkin Humana Press Inc., Totowa, NJ

The rationale for, and effective use of, studies utilizing nonconcurrent (historical) controls and observational research for cardiovascular and orthopedic products is described, in addition to the examples of randomized controlled trials. In the chapters on heart valves and total hip replacements a brief history serves to emphasize that incremental changes to device design and surgical techniques occur constantly over time, and that the impact of such incremental changes are unlikely to be detected in a comparative prospective pivotal trial. Limitations imposed by subject sample sizes, trial duration, and lack of sensitivity in relevant clinical endpoints all contribute to the use of alternative research methods to supplement information gathered from clinical experience. Hence, as these authors note, there is and should be a continued reliance on extensive nonclinical research (bench testing and animal models) in the evaluation of device performance.

Postmarket surveillance studies, critical to the evaluation of long-term device performance, failure investigations, and unanticipated safety issues, are represented in Chapter 7, Chapter 8, Chapter 11, and Chapter 12. Logistical difficulties usually preclude the prospective tracking of patients beyond a 2-yr trial, yet gathering data on long-term clinical performance is critical to good patient care and the development of improved products. These chapters illustrate the alternatives to long-term prospective trials—observational studies utilizing appropriate statistical analysis plans and controls, well-considered explant retrieval and analysis studies, and hypothesis-driven failure analysis investigations that include bench testing, as well as in vivo and in vitro nonclinical research. The examples included in these chapters illustrate that ultimately no amount of nonclinical research can substitute for, or truly predict, clinical experience with a medical device, and yet clinical experience cannot be understood without the support of subsequent nonclinical research studies. Elucidating pacemaker-lead failure modes postmarketing is a now classic example of this research challenge (Chapter 12). The possible role of implanted products in causing systemic disease, notably silicone gel-filled breast implants and autoimmune disease, is a more recent example of a long-term safety issue that could not have been predicted, or reliably addressed in nonclinical studies, prior to marketing.[1] In these situations, both clinical and nonclinical research are necessary to resolve the issue once suggestive data from clinical experience is in hand. The satisfactory resolution of a similar controversy with injectable collagen, based on supplemental clinical research coupled with a weight of the evidence evaluation of all the available nonclinical and clinical data, is described in Chapter 8.

The thoughtful approaches to experimental and observational studies of devices developed by the investigators in the following case studies demonstrate that clinical research on medical devices can, and routinely should, reflect the level of scientific rigor we have come to take for granted in studies of pharmaceuticals.

Reference

1. Rose, N. R. 1996. The silicone breast implant controversy: The other courtroom. *Arthritis Rheumatism* 39:1615–1618.

5

Clinical Studies of Prosthetic Heart Valves Using Historical Controls

Gary L. Grunkemeier

"Quantitative, comparative clinical trials can sometimes be better accomplished with technics other than randomization for selection of a control group. These include selection of literature controls, matched controls and controls from a previous study."[1]

"For both statistical and ethical reasons, adaptive designs should be used more often, and strictly randomized designs should be used sparingly."[2]

"Whether we like it or not, most of our future decisions about medical practice, health care, and scientific technology will have to be made without evidence from randomized trials...."[3]

1. Introduction

Beginning in 1960, replacement of diseased heart valves has dramatically lowered the death rate and improved the functional class of patients compared with the natural history of heart valve disease. Experience has proven that acceptable replacement devices for diseased heart valves can be produced and evaluated clinically in single-arm studies, without using randomized or other concurrent controls. The elements of this successful development have been thoughtful design and engineering, carefully controlled manufacturing specifications,

From: *Clinical Evaluation of Medical Devices: Principles and Case Studies*
Edited by K. B. Witkin Humana Press Inc., Totowa, NJ

thorough laboratory pulse-duplicator and wear testing, animal implantation, and limited initial clinical use with thorough and complete hemodynamic and clinical follow-up.

This chapter will discuss the historical perspective pertaining to the first successful heart valve and the ensuing development and present availability of replacement valves (the majority of this development took place in a relatively unregulated environment), some theoretical and practical arguments against the use of randomized clinical trials (RCT) to evaluate medical devices, especially heart valves, and the federal government regulatory guidelines, including the recent efforts leading to a revised heart valve guidance document, that resulted in the use of literature controls to set criteria for clinical studies supporting marketing approvals.

2. Historical Background

The first clinically successful valve was the Starr-Edwards caged-ball valve.[4] It is still in use today, with only a slight modification of the original design.

2.1. The Starr-Edwards Story

Inventor Lowell Edwards and cardiac surgeon Albert Starr began their collaboration in 1958.[5] Many investigators had been experimenting with heart valve designs,[6-8] but no prosthesis was widely available to replace diseased heart valves. Some details of this collaboration are of interest to illustrate how research experimentation can progress successfully using strategies that do not involve randomization.

A heart valve functions as a simple check valve, allowing blood to flow in a forward direction, while preventing retrograde flow. After evaluating several types of mechanisms, Starr and Edwards decided on a simple caged-ball configuration, in which the ball opened and closed the orifice by moving to and fro within the confines of a small cage. After refining this design until it worked satisfactorily in the bench-testing environment, they began implanting the valves into dogs. Dogs have increased blood clotting activity compared to humans, and not surprisingly, most of the early implants clotted shut in the early postoperative period. Clots organized around the suture line and grew out on the inflow side of the valve, obstructing the motion of the ball.

To be able to study the long-term effects of this unnatural device in the heart, they produced a modified version of the valve with a sili-

Fig. 1. The original Starr-Edwards caged ball valves. Shielded model **(left)**, which was successful in the dog, and unshielded model **(right)**, which was used in the first patient.

cone rubber shield that snapped down over the suture line. This prevented blood contact with the suture line and resulted in long-term canine survivors.

2.1.1. The First Valve Selection Dilemma

Now Starr and Edwards were ready for the first human implants, and they had two versions of the valve: the shielded version (Fig. 1, left), which worked well in dogs, and the simpler, nonshielded valve (Fig. 1, right), which did not. Some might argue that the proper approach would have been to begin a randomized study of these two designs. However, Starr and Edwards reasoned that since humans are much less prone than dogs to develop clots on foreign substances, the shield would not be needed. They decided to use the simpler valve, with less bulk and less complexity to the surgical procedure. They would reserve the option of switching to the shielded valve if the unshielded version did not perform adequately. The first patient died shortly after the operation, but the second patient was rehabilitated and lived for 10 yr with the new valve, dying in an accidental fall.

2.1.2. Prospective Clinical Database

Dr. Starr decided that introduction of a new therapy should include an obligation to monitor all patients' progress for their lifetimes, so he began a follow-up effort that relied on clinic visits, telephone contacts, and periodic questionnaires. Within a short time, examination of the resulting follow-up records revealed that the major postoperative complications were cerebral events caused by thromboemboli, tiny blood clots that formed at the cloth–metal interface on the inflow

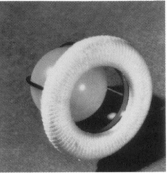

Fig. 2. Original production model of the Starr-Edwards valve **(left)**, used from 1960 to 1965, and modified version **(right)** used from 1965 to the present.

orifice of the valve and subsequently migrated into the bloodstream, sometimes lodging in the brain. In 1965, a subtle but important modification was made to the original valve in response to this complication. The cloth of the sewing ring was extended to cover the complete inflow surface of the housing. This placed the interface with the metal in the fastest part of the flow, where such clots were less likely to develop.

2.1.3. The Second Valve Selection Dilemma

Once again there were two valves available: the original valve (the housing had been changed from plastic to metal) with much exposed metal (Fig. 2, left), and the modified version with no exposed metal (Fig. 2, right). Starr and Edwards followed their clinical insight and switched all implanted devices to the new model, rather than performing a randomized comparison. They continued to follow patients with both valve models, and it was soon obvious that a dramatic reduction in thromboembolic episodes had been achieved (Fig. 3). The modified valve is now in its 32nd year of clinical use.

2.2. Development of Other Heart Valve Prostheses

Alternative valve designs underwent rapid development during the 1960s and 1970s, usually originating from a collaboration between cardiac surgeons and biomedical engineers. Progress often began when a member of the team envisioned a new design or an improvement to an existing design; they optimized the concept in the laboratory, implanted the resulting device into laboratory animals, and then, if acceptable, into humans.

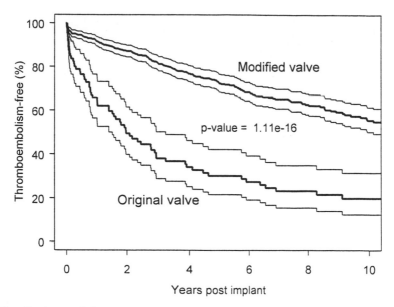

Fig. 3. Actuarial curves (Kaplan-Meier method) for the freedom from thromboembolism for two heart valves available in 1965 (*see text*). The thinner lines show 95% confidence intervals. The *p*-value, in scientific notation, indicates an extremely significant difference.

2.2.1. Successful Valves

From this process, many successful valves have been developed. These have included very different looking designs, some mechanical and some made from biological tissue. The mechanical designs have included single disk valves (Fig. 4, left) and bileaflet valves (Fig. 4, right). The biological valves have included those incorporating porcine aortic valves (Fig. 5, left) and those tailored from bovine pericardium (Fig. 5, right). Through 1994, over 2 million valves had been implanted, approx 60% of them of mechanical design and 40% biological (refer to Table 1).[9] It is widely acknowledged that none of these valves are yet perfect, but all of the valves in current use are considered clinically acceptable.

2.2.2. Unsuccessful Valves

Some valves were tried clinically but discontinued when their performance was found to be inferior to the contemporary standards. Complications of heart valves are relatively easy to diagnose, and inferior designs are usually discarded after limited clinical use.

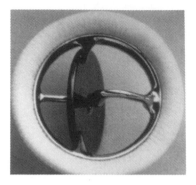
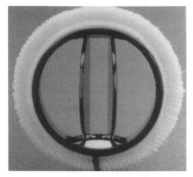

Fig. 4. Other current mechanical valves: tilting disk **(left)** and bileaflet valve **(right)**.

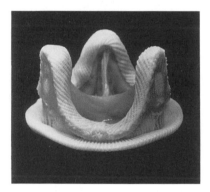

Fig. 5. Heart valves of biological origin: porcine **(left)** and bovine pericardial **(right)**.

In spite of extensive development and thorough preclinical testing, occasionally a valve will be considered for human implantation that has a higher than acceptable complication of some unexpected type. The valve is released for limited clinical testing, and the problem is usually discovered within a year or so and after several hundred implants. The implantation is then stopped, while the problem is exhaustively analyzed using bioengineering techniques, and the patients who had already received the valve are closely monitored.

2.2.3. Government Regulation

The heart valve review process was originally informal, depending on feedback from physicians who collected data by following their patients' progress. These physicians reported problems back to the

Table 1
Type, Design, Manufacturer, and Approximate Numbers of Implants
(In Thousands) of Heart Valves in Use Through 1994[9]

Type	Design	Manufacturer	No. of implants ($\times 1000$)
Mechanical	Ball	Baxter Edwards	157
	Disk	Medtronic	178
		Medical Inc.	60
		Sorin	212
	Bileaflet	St. Jude	580
		Baxter Edwards	32
		CarboMedics	110
		Sorin	8
		ATS	5
Biological	Porcine	Medtronic	288
		Baxter Edwards	466
		St. Jude	28
	Pericardial	Baxter Edwards	35
		Mitral Medical	13
		Sorin	8
	Homograft	Cryolife	14

manufacturers, and to their peers at society meetings and through journal articles. Then, in 1976, the Food and Drug Administration (FDA) was given the added responsibility for ensuring and monitoring the safety and efficacy of medical devices and began an evolutionary process of expanding requirements for approval, which continues today.

Altogether, these informal and formal restraints have been very successful. Only a very few heart valves with unacceptable failure rates have progressed to the stage of wide clinical use. The most notorious example is a tilting disk valve that was implanted from 1978 to 1986 in approx 86,000 patients.[10] Its failures have been widely publicized in the lay press; however, fewer than 1% of these disk valves are known to have failed to date, and about a third of the patients experiencing failure were saved by emergency reoperation.[11]

The FDA periodically produces documents containing information to guide premarket approval applications (PMA). By 1992, the FDA guidance document for heart valves was quite rigorous with regard to the collection of information concerning patient condition and valve complications, but required a rather small clinical study (a minimum of 35 aortic and 35 mitral valves from each of three centers followed for 1 yr) based only on subjective opinion, and did not require any formal evaluation or comparison of the resulting valve performance measurements.

3. Randomization in Heart Valve Studies

In 1992, a report commissioned by Congress, the "Final Report of the Committee for Clinical Review" (also known as the Temple Report)[12] criticized the FDA device approval process for having far less rigorous requirements than the traditional drug approval process. As a result, the FDA decided to use heart valves as a pilot device to increase the scientific rigor, and in 1993 issued a draft document containing suggested revisions to the requirements for heart-valve testing. Patterned after the historically successful drug approval studies, this document suggested that an RCT be required for heart valves.

3.1. Problems with Randomization of Medical Devices

A prospective RCT is considered necessary to establish cause–effect relationships in clinical studies; that is, to demonstrate that an observed difference between groups receiving different treatments is caused by the treatment. Random allocation of patients provides unbiased estimates of treatment effects and valid probability statements for hypothesis tests in such comparative studies. However, there are compelling reasons for using alternative methods to evaluate medical devices, especially artificial heart valves.[13]

3.2. Devices Are Not Drugs

Random allocation of patients is considered necessary to establish efficacy in drug evaluation studies, since the effect is nebulous and small relative to the biological variability, especially if the effect is difficult to measure objectively. Ideally, blinding (masking) is invoked to eliminate potential placebo effect. Finally, a very large number of patients are often needed, since the signal-to-noise ratio is low.

Major differences between drugs and implanted medical devices must be addressed before designing an appropriate RCT. For drugs, the early clinical phases of testing are designed to establish a range of safe dosage levels, and within that safe range, the most effective dose. Then, the RCT is designed to establish (hopefully) superior efficacy to some control drug. This can be difficult because the effect of a drug in humans is often difficult to predict from preclinical studies, the exact mechanism of which may be unknown, and there is usually a wide range of individual sensitivity to its action because of biological variability. However, the effectiveness of a medical device (its adequacy to perform its mechanical function) can be tested using in vitro studies and in animals, and can be measured directly in humans. In the case of a heart valve, this function can be measured by means of such parameters as gradient, cardiac output, transvalvular regurgitation, and so forth, and can even be inferred to a large extent by a lack of symptoms.

3.3. Unattainable Sample Sizes

The definitive objective for the clinical study of a medical device is to demonstrate relative safety. Thus, for a heart valve, the study must demonstrate that the complications attendant to the use of the new valve are not significantly higher than with other currently acceptable valves. Since the major complications with current valves are low, a large number of patients would be needed in a randomized study. To achieve the same statistical significance and power as using historical controls, approximately four times as many patients would be needed.[1] Moreover, the existence of a randomized study interferes with normal patient referrals, which may reduce the number available for the study. Patients needing heart valve replacement are not nearly as plentiful as those suffering from headache, high blood pressure, high serum cholesterol, and so forth, who could be recruited for a drug study. One estimate of the number of patients needed for a single heart valve RCT was equal to one-third of all valves implanted in the United States in 1 yr.

3.4. Ethical Problems

In experiments in medicine, and especially in surgery,[14-17] randomly allocating subjects to therapy introduces complications that may make it more harmful than helpful. One difficulty with randomized surgical studies is that a predefined static protocol is not responsive to changes as surgical skill evolves and experience regarding patient selection is acquired. In particular with heart valves, there is an ethical problem in that two heart valves are rarely regarded as equal for a given patient. Instead, performance characteristics of the valve are matched to the particular clinical situation of each patient.[18,19]

Randomization eliminates the clinical judgment of the physician and also interferes with subtle aspects of the patient–physician relationship.

3.5. Rare Events

Most complications associated with heart valves, such as thromboembolism and bleeding, are related to patient susceptibility factors, and so far seem to be unavoidable. An exception is structural failure of a mechanical valve, which is a most dreaded, unacceptable complication. When such a failure mode does occur, it is usually so rare that the RCT would not be large enough to detect it. Thus, continued postmarket surveillance is necessary.

3.6. Late Events

Structural failure for biological (or tissue) valves, on the other hand, is inevitable. Tissue valves generally have good performance profiles except for structural failure, which is the fundamental complication that continues to drive the development of tissue valve technology. The failure rate increases with time from implantation, but current tissue valves have failure rates of only 10–20% at 10 yr. To detect a statistically significant change would require a huge RCT, not only in terms of patient-years but also in elapsed time, to beyond 10 yr or so. Obviously, the resources to do such studies for premarket approval are not available, and if they were, by the time an approval was given, the valve would already be outdated.

3.7. External Versus Internal Validity

Random allocation of patients ensures *internal* validity within a study, that is, an unbiased comparison between the two treatments. However, randomized studies have narrow inclusion criteria for patients. In addition, only certain physicians will want to conduct such a study, only certain types of (eligible) patients will agree to participate, and the fact that a study is being performed at a center may further limit the types of patients who are referred there for treatment. Thus, the results may not be applicable to most patients, surgeons, and centers in which the new treatment will be used after its market release. That is, the study will not have *external* validity.[20, 21]

The purpose of the clinical PMA study should be to maximize the ability to show, with a reasonable amount of time and effort, that the expected complication rate with the new valve is within the clinically

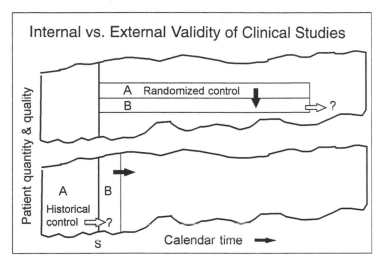

Fig. 6. Schematic of two studies comparing an old (**A**) and new (**B**) heart valve in two identical groups of patients. The patient groups are shown by the closed irregular shapes: The horizontal dimension is time and the vertical dimension represents patient profiles that change, irregularly, over time. At a point in time shown by the vertical lines at S, clinical comparison of the new valve is begun. In the upper diagram, patients are randomly allocated to valve A (old, control) or B (new, treatment). The two groups are thus similar in number and patient quality, indicated by the identical regular rectangles. This comparison has internal validity, shown by the filled arrow. However, at the end of the study, the right end of the rectangles, valve B may now be used in the general population. The external validity is not assured, as indicated by the open arrow with the question mark. In the lower diagram, historical information about valve A is used for comparison. The internal validity is not assured, as shown by the arrow followed by a question mark. The study is concluded much quicker, since valve A does not have to be re-evaluated as part of the study, and thus fewer total patients are needed. Since valve B is used in virtually all patients who would qualify, and it will (on approval) be used in a similar group of patients, the external validity is improved as indicated by the solid arrow. The chronological design illustrated in the lower diagram is traditionally depended on to estimate event rates, where randomization is impossible. For example, accident rates, crime rates, divorce rates, and so forth, are tracked, and comparisons to previous times are made to assess changes.

acceptable range of current valves. Thus, all patients who fit the criteria and who consent should be included in the study in order to estimate the parameters of the new valve. This concept of external vs internal validity is illustrated in Fig. 6.

3.8. Published Randomized Valve Studies

For reasons given above, we would expect that randomized studies of heart valves are uncommon and produce limited practical information in a time-delayed fashion. A comprehensive review[22] of hundreds of published studies of clinical valve performance revealed eight that were randomized, including the two best-known ones.[23, 24] These studies were all of rather small size, about 50–200 valves in any position. They encountered design problems, including pooling different positions and models of valves and changing the models used during the course of the study.

The results of these studies were usually negative (i.e., no significant differences), which may be a result of limited sample size and duration of follow-up. Where there has been a positive finding, it confirmed what was known previously from observational studies; biological valves have more structural failure and mechanical valves have more bleeding complications. In many cases, by the time the results of the study were published, the valves being studied were no longer being sold. This review concluded that most of the information available about valve performance still has to be gleaned from high-quality studies involving a large number of patients (observational studies, databases) as opposed to randomized studies. Information about the ultimate performance characteristics, especially regarding very low-risk complications, can also be obtained from a valve manufacturer's implant registry by incorporating assumptions about patient decrement resulting from mortality and underreporting of events.[25]

4. FDA Guidance Document Revision

Prompted in part by the Temple Report,[12] artificial heart valves were among the first devices to undergo comprehensive restructuring of their PMA guidance document. In considering how to upgrade these requirements, the draft guidance issued by FDA suggested that the clinical study should be based on a two-tailed hypothesis test at the 5% significance level, with 95% power to detect a doubling of the 1-yr complication rates with the study valve, compared to the control. Randomized controls were declared not mandatory, but would be "appropriate and beneficial."

4.1. Previously Proposed Study Design

During 1993, FDA actively sought responses to this proposal. Several meetings were held to discuss the proposed guidance document, including workshops on in vitro testing and clinical trial design.[26] As a result of this interaction among FDA, manufacturers, cardiologists, cardiac surgeons, and other interested persons, a consensus developed and became the basis for the clinical study section of the revised guidance document, issued late in 1993.[27]

The requirement of randomization was deemed not practicable for heart valves. The optimal study design was finally selected by the process of elimination. Without randomization, there could be concurrent controls, but selection bias would undoubtedly occur, and the choice of a control valve was problematic. Therefore, the decision was made to use objective criteria based on previous studies (historical controls). The study design that was chosen had been suggested in a 1986 publication by a team of physicians and scientists who were experts in the use of heart valves and the evaluation of heart valve data.[28] This publication, after thoroughly discussing the issues, recommended an approach that would compare the results with a new valve to average event rates taken from the literature, as a basis for approval of new valves.

4.2. Objective Performance Criteria

One reason that objective criteria could be considered in this situation was that the complications observed during evaluation of heart valves and the definitions of these complications had been previously determined by a joint committee of the American Association for Thoracic Surgery and the Society of Thoracic Surgeons, and simultaneously published by three major cardiac journals.[29-31] These guidelines were coauthored by five experienced cardiac surgeons, with collaboration and input from an illustrious team of surgeons and cardiologists* with vast experience in heart valve research and in caring for and following heart valve patients.

The complications used as objective criteria are structural deterioration, nonstructural dysfunction, thromboembolism and thrombosis,

---

* The team of physicians consisted of C. W. Akins, L. H. Burr, D. M. Cosgrove, A. R. C. Dobell, P. A. Ebert, B. J. Gersh, W. R. E. Jamieson, J. W. Kirklin, N. T. Kouchoukos, H. Laks, F. D. Loop, D. J. Magilligan, Jr., D. G. Pennington, D. N. Ross, H. V. Schaff, and A. S. Wechsler.

Table 2
Definitions of Morbidity from Guidelines for Reporting[29-31]

Event	Includes	Excludes
Structural deterioration	Deterioration, wear, stress fracture, poppet escape, calcification, leaflet tear, stent creep	Infection, thrombosis
Nonstructural dysfunction	Entrapment by pannus or suture, inappropriate sizing, hemolytic anemia leak	Thromboembolism, infection
Valve thrombosis	Thrombosis proved by operation, autopsy, or clinical investigation	Infection
Thromboembolism	Neurological deficit, peripheral arterial emboli, acute myocardial infarction (in patients with normal coronaries or those < 40 yr old)	Septic emboli, hemorrhage, surgical events
Anticoagulant bleed	Bleeds causing death, stroke, operation, hospitalization or transfusion in patients receiving anticoagulant or antiplatelet drugs	
Endocarditis	Based on blood cultures, clinical signs, and/or histological evidence at reoperation or autopsy	

anticoagulant-related hemorrhage, and prosthetic valve endocarditis. These complications were carefully defined (*see* Table 2), and the guidance document containing them was widely accepted and had been used by many authors, describing the results with approved, acceptable heart valves. Thus, there was a published basis for defining the objective standards for a new valve. Much of this literature had been recently reviewed and summarized (*see* Table 3).[22]

FDA officials reviewed this literature, using strict criteria, and added information from approved PMA valve submissions. FDAs review resulted in 45,000 patient-years of data from 10,000 patients with currently marketed valves from which the final criteria were derived. It was found that the current valve models had similar average rates (independent of specific models) but differed slightly between mechanical and biological valves, and that the range for each complication was from approx 0 to about twice the average complication rate. The term objective performance criteria (OPC) was coined to designate the average event rates for these critical complications. The resulting OPC were provided separately for mechanical and biological valves (*see* Table 4).

Table 3
Results of Recent Literature Review of Heart Valve Performance[22]

Valve type	Model	Series	Patients	Patient-years
Mechanical	Starr-Edwards	10	8224	41,165
	Björk-Shiley Standard	14	4203	26,769
	Bjork-Shiley CC/Monostrut	7	4271	9036
	Medtronic Hall	6	5360	20,311
	OmniScience	4	919	2055
	St. Jude Medical	13	6717	17,003
Biological	Hancock	17	8216	30,546
	Carpentier-Edwards	12	5942	32,430
Total		83	43,852	179,315

Table 4
OPC from FDA Guidance Document[a]

Complication	Mechanical	Biological
Thromboembolism	3.0	2.5
Thrombosis	0.8	0.2
All bleed	3.5	1.4
Major bleed	1.5	0.9
All leak	1.2	1.2
Major leak	0.6	0.6
Endocarditis	1.2	1.2

[a]The numbers are "linearized" rates (constant hazard functions) that have units of events per 100 patient-years, or percent per year. See ref. *27*.

4.3. Sample Size Requirements

The final guidance document, released in December 1993,[27] incorporated a strategy very similar to that advocated by Gersh et al.[28] A new valve must have significantly lower complication rates, statistically, than two times the OPC. A decision was made to use OPC in the range of 1.2%/yr and above as the minimum for purposes of sample size estimation, since requiring a new valve to meet the OPC condition for every complication listed would require too large a sample. The statistical requirements were declared in hypothesis-testing terminology (a one-sided hypothesis test at the 95% level with 80% power). The (null) hypothesis to be rejected is that the new valve does have at least twice the OPC complication rate. It was shown that about 800 patient-years of follow-up would be required to show that

Table 5
In Vitro Testing Requirements for PMA Studies[33]

1. Identify cyclic failure mode.
2. Complete finite element analysis (FEA) to identify regions of high stress.
3. Conduct wear studies to identify regions of high wear and to validate FEA.
4. Determine fatigue characteristics of the structural components.
5. Calculate life of the structural components under physiological conditions.
6. Establish appropriate inspection methods based on fatigue characteristics.

a new valve met this condition.[32] There were some further requirements for this minimum follow-up. The 800 patient-years should be equally divided between positions; i.e., there should be a minimum of 400 aortic and 400 mitral valves. Also, there should be a minimum of three centers and 50 patients in each position at each center; these 300 patients must be followed for at least 1 yr. Finally, complete follow-up is required on a minimum of 15 patients of each size and position, also followed for a minimum of 1 yr.

4.4. FDA Science Forum Presentation

A compact review of the new guidance document requirements was contained in a poster presentation exhibited at an FDA Science Forum on Regulatory Sciences in 1994.[33] The senior author was the FDA official primarily responsible for the final version of the draft guidance document. In addition to the above information on OPC, this presentation contained information on other features and requirements of the new guidance document. For example, the in vitro testing requirements include several carefully defined steps (*see* Table 5). The required endpoints for the clinical study went well beyond the specific complications discussed above, and include the many endpoints needed to determine safety, effectiveness, and clinical utility (*see* Table 6). The conclusions from this poster presentation (*see* Table 7) were very practical, and incorporated many of the points discussed above.

4.5. Future of the FDA Guidance Document

The 1993 FDA guidance document for heart valve studies greatly augmented the scientific merit of the PMA approval process compared to previous guidance documents. However, this is not a static document. One year after the first version was issued, a slightly revised version was issued with minor changes in grammar and semantics.[34]

Table 6
Endpoints Required by FDA Guidance Document
to Determine Safety, Effectiveness, and Clinical Utility[33]

Morbid event rates	Thromboembolism, thrombosis, hemorrhage, endocarditis, paravalvular leak, structural deterioration, angina, arrhythmia, cardiac arrest, heart failure, hemolysis, myocardial infarction, nonstructural dysfunction
Rates for related consequences	Death, explant, reoperation
Hemodynamic performance	Low gradient across open valve, leakage across closed valve
Blood studies	Hemoglobin, serum LDH, haptoglobin, reticulocytes—with trend analyses to determine increases over time—and RBC, WBC, hematocrit
Functional classification	New York Heart Association functional class to assess quality of life

Table 7
Conclusions from the Poster Presentation[33]

1. Extensive in vitro testing must precede clinical studies.
2. RCTs are not feasible for several reasons, including the large sample sizes required and the difficulty in establishing a control valve.
3. Fixed standards (OPC) are possible because of established definitions used in the published literature.
4. No short-term study can serve as an early warning for late events. Post-approval follow-up is necessary to address long-term issues.

The document itself contains provisions for a review every 3–5 yr to determine if adjustments are needed. It may also move toward other areas, such as quantifying the effects of patient risk factors and incorporating actuarial rates into the OPC, rather than simple linearized rates. It will also consider establishing criteria for other clinical endpoints, e.g., blood pressure drop and regurgitation.

5. Conclusion

Heart valve development over the past 30 yr has progressed by using careful preclinical and clinical studies to evaluate valve performance. These valves have all been established using the same basic model:

1. Engineering development of a new design with theoretical improvements, for example in flow characteristics;
2. Extensive in vitro testing, including pulse duplication studies and wear testing;
3. Animal implants;
4. Careful, limited, single-arm clinical investigation with detailed follow-up of results;
5. Open marketing after successful comparison of results with a growing compendium of information on previous usage of currently acceptable devices; and
6. Continued follow-up of the device, with the ultimate test being the crucible of clinical scrutiny.

The single-arm aspect of the definitive clinical studies has been possible because the major heart valve complications are relatively objective and easy to diagnose, the complications used as objective valve performance criteria have been determined and the definitions have been standardized, and these definitions have been widely accepted and used for several years so there is a body of information available for currently available heart valves.

Randomized allocation of treatment has valuable practical application in many settings, but for initial device evaluations, observational studies are more appropriate. Trying to preserve the theoretical integrity of a study at all costs, by forced randomization when it is not practicable, introduces bias and other problems, and requires more patients than are available for heart valve studies. Randomized studies of approved valves have only confirmed relationships that were previously known from observational studies. Marketing approval can be based on an absolute comparison with qualifying standards through the use of historical controls. However, short-term studies (randomized or not) can not detect increased rates of rare or late events. Long-term postmarket studies, including RCTs where appropriate, can be used to detect rare complications, and differences with regard to late-occurring complications among otherwise comparable valves.

References

1. Gehan, E. A. and Freireich, E. J. 1974. Non-randomized controls in cancer clinical trials. *N. Engl. J. Med.* 290:198–203.
2. Weinstein, M. C. 1974. Allocation of subjects in medical experiments. *N. Engl. J. Med.* 291:1278–1285.

3. Feinstein, A. R. 1984. Current problems and future challenges in randomized clinical trials. *Circulation* 70:767–774.
4. Starr, A. and Edwards, M. L. 1961. Mitral replacement: clinical experience with a ball-valve prosthesis. *Ann. Surg.* 154:726–740.
5. Lefrak, E. A. and Starr, A. 1979. Starr-Edwards ball valve. In *Cardiac Valve Prostheses.* Appleton-Century-Crofts, New York, Ch. 3.
6. Lillehei, C. W., Barnard, C. N., Long, D. M., Jr., Schimert, G., and Varco, R. L. 1961. Aortic valve reconstruction and replacement by total valve prostheses. In *Prosthetic Valves for Cardiac Surgery* (Merendino K. A. ed.). Thomas, Springfield, pp. 527–575.
7. Braunwald, N. S., Cooper, T., and Morrow, A. G. 1960. Complete replacement of the mitral valve: successful clinical application of a flexible polyurethane prosthesis. *J. Thorac. Cardiovasc. Surg.* 40:1–11.
8. Harken, D. E., Soroff, H. S., Taylor, W. J., Lefemine, A. A., Gupta, S. K., Lunzer, S., and Low, H. B. C. 1961. Aortic valve replacement. In *Prosthetic Valves for Cardiac Surgery* (Merendino K. A. ed.). Thomas, Springfield, pp. 508–526.
9. Grunkemeier, G. L., Starr, A., and Rahimtoola, S. H. 1996. Prosthetic heart valves. In *Hurst's The Heart Update I* (O'Rourke R. A. ed.). McGraw-Hill, New York, pp. 95–123.
10. Hiratzka, L. F., Kouchoukos, N. T., Grunkemeier, G. L., Miller, D. C., Scully, H. E., and Wechsler, A. S. 1988. Outlet strut fracture of the Björk-Shiley 60° Convexo-Concave valve: current information and recommendations for patient care. *J. Am. Coll. Cardiol.* 11:1130–1137.
11. Grunkemeier, G. L. and Anderson, W. N., Jr. 1996. Passive surveillance of heart valve devices: Björk-Shiley outlet strut fracture rates. Longterm effects of medical implants. 5:155–168.
12. Food and Drug Administration. 1993. *Final Report of the Committee for Clinical Review. "The Temple Report."* FDA Report. March 1993, pp. 1–45.
13. Grunkemeier, G. L. and Starr, A. 1992. Alternatives to randomization in surgical studies. *J. Heart Valve Dis.* 1:142–151.
14. Love, J. W. and Phil, D. 1975. Drugs and operations: some important differences. *JAMA* 232:37–38.
15. Bonchek, L. I. 1979. Sounding board: are randomized trials appropriate for evaluating new operations? *N. Engl. J. Med.* 301:44–45.
16. Anderson, R. P. 1980. Standards for surgical trials. *Ann. Thorac. Surg.* 29:192.
17. van der Linden, W. 1980. Pitfalls in randomized surgical trials. *Surgery* 87:258–2262.
18. Starr, A. and Grunkemeier, G. L. 1988. Selection of a prosthesis for aortic valve replacement. *Eur. Heart J.* 9-E:129–137.
19. Starr, A. and Grunkemeier, G. L. 1994. Selecting a prosthetic valve. In *Current Therapy in Cardiovascular Disease* (Hurst J. W. ed.). 4th ed. Mosby, St. Louis, pp. 236–241.
20. Kramer, M. S. and Shapiro, S. H. 1984. Scientific challenges in the application of randomized trials. *JAMA* 252:2739–2745.

21. Olschewski, M., Schumacher, M., and Davis, K. B. 1992. Analysis of randomized and nonrandomized patients in clinical trials using the comprehensive cohort follow-up study design. *Control. Clin. Trials* 13: 226–239.

22. Grunkemeier, G. L., Starr, A., and Rahimtoola, S. H. 1992. Current problems in cardiology. *Mosby Year Book* 17:335–406.

23. Bloomfield, P., Wheatley, D. J., Prescott, R. J., and Miller, H. C. 1991. Twelve-year comparison of a Björk-Shiley mechanical heart valve with porcine bioprostheses. *N. Engl. J. Med.* 324:573–579.

24. Hammermeister, K. E., Sethi, G.K., Henderson, W.G., Oprian, C., Kim, T., and Rahimtoola, S. 1993. A comparison of outcomes in men 11 years after heart-valve replacement with a mechanical valve or bioprosthesis. *N. Engl. J. Med.* 328:1289–1296.

25. Grunkemeier, G. L., Chandler, J. G., Miller, D. C., Jamieson, W. R. E., and Starr, A. 1993. Utilization of manufacturer's implant card data to estimate heart valve failure rates. *J. Heart Valve Dis.* 2:493–503.

26. Grunkemeier, G. L. 1993. Will randomized trials detect random valve failure? Reflections on a recent FDA workshop. *J. Heart Valve Dis.* 2: 424–429.

27. Prosthetic Devices Branch, Division of Cardiovascular, Respiratory and Neurological Devices. 1993. Draft Replacement Heart Valve Guidance. December 7.

28. Gersh, B. J., Fisher, L. D., Schaff, H. V., Rahimtoola, S. H., Reeder, G. S., Frater, R. W. M., and McGoon, D. C. 1986. Issues concerning the clinical evaluation of new prosthetic heart valves. *J. Thorac. Cardiovasc. Surg.* 91:460–466.

29. Edmunds, L. H., Jr., Clark, R. E., Cohn, L. H., Miller, D. C., and Weisel, R. D. 1988. Guidelines for reporting morbidity and mortality after cardiac valvular operations. *Ann. Thorac. Surg.* 46:257–259.

30. Edmunds, L. H., Jr., Clark, R. E., Cohn, L. H., Miller, D. C., and Weisel, R. D. 1988. Guidelines for reporting morbidity and mortality after cardiac valvular operations. *J. Thorac. Cardiovasc. Surg.* 96: 351–353.

31. Clark, R. E., Edmunds, L. H., Jr., Cohn, L. H., Miller, D. C., and Weisel, R. D. 1988. Guidelines for reporting morbidity and mortality after cardiac valvular operations. *Eur. J. Cardiothorac. Surg.* 2:293–295.

32. Grunkemeier, G. L., Johnson, D., and Naftel, D. C. 1994. Sample size requirements for studying heart valves with constant risk events. *J. Heart Valve Dis.* 3:53–58.

33. Johnson, D. M., Naftel, D. C., and Grunkemeier, G. L. 1994. Premarket evaluation of heart valves: a prospective non–randomized study. Poster presentation at the FDA Science Forum on Regulatory Sciences, September 29, Washington, DC.

34. Division of Cardiovascular, Respiratory and Neurological Devices. 1994. Draft Replacement Heart Valve Guidance. October 14.

6

Prospective Multicenter Clinical Trials in Orthopedics
Special Concerns and Challenges

John D. Van Vleet

1. Introduction

The clinical evaluation of orthopedic devices is, in the overall scheme of clinical research, a relatively recent phenomenon. Although innovations in orthopedics date back over 30 yr and the 1976 Medical Device Amendments of the Food, Drug, and Cosmetic Act are commemorating 20 yr, it has been scarcely more than a decade that orthopedic clinical studies have been conducted with any degree of consistency. Pharmaceutical clinical research has a more lengthy history and, as such, has been frequently—and at times inappropriately—used as a template for establishing standards for the conduct of orthopedic research. Both, however, share certain basic elements that are central to realizing good clinical research.

To fully understand the issues involved in the investigation of orthopedic devices, it is important to examine the industry from a historical perspective. The following discussion briefly summarizes the history of orthopedic clinical research. Although it does not provide a comprehensive listing of all regulatory requirements, the requirements with unique applications to orthopedic research are discussed, which necessitates delving into the nuances of orthopedic clinical practice and regulatory intent.

From: *Clinical Evaluation of Medical Devices: Principles and Case Studies*
Edited by K. B. Witkin Humana Press Inc., Totowa, NJ

2. History of Orthopedic Clinical Research

In 1895, a Frenchman named Revra DePuy introduced the world of orthopedic medicine to the innovation of limb-molded, wire-mesh alternatives to bulky splints and casts. For almost 70 yr, orthopedic devices were limited to mere variations on the theme of "soft goods"; in other words, those involved in the management of fractures, sprains, congenital malalignments, and other orthopedic maladies. This collection of devices provided noninvasive treatment for the most common orthopedic problems, mainly those associated with trauma.

A routine (and to this point untreatable) occurrence in the elderly population was fracture of the femoral neck. This type of fracture, coupled with the frequent incidence of osteoarthritis in the hip joint and its painful sequelae, prompted surgical investigation in the field of joint resurfacing, or arthroplasty. For several years, widespread experimentation in the orthopedic community was minimally successful to disastrous. Then, in 1962, John Charnley, an English orthopedist, perfected a technique involving the implantation of a ball-tipped stainless-steel rod into the proximal femur that was designed to articulate into a polyethylene hip cup cemented in the acetabulum.[1] This procedure, developed and mass-produced in cooperation with the English company C. F. Thackray, resulted in the subsequent knighting of Charnley (*see* Fig. 1). Charnley's technique has remained essentially unchanged as the gold standard on which each succeeding arthroplasty system has been based or to which it is compared. Its use has been successfully extrapolated from the hip joint to other diseased articulating surfaces, such as the knee and the shoulder. DePuy's company acquired C. F. Thackray in 1990, merging the oldest orthopedic institution in the world with the company responsible for bringing the orthopedic world into the 20th century.

3. Classification of Devices

As orthopedic medicine advanced technologically and grew in complexity, a host of devices and concepts found their way into general practice. In an effort to manage this increasingly complex field, the US Food and Drug Administration (FDA) attempted to classify devices based on the risk posed to the end recipient.

Fig. 1. Sir John Charnley.

Volume 21 of the Code of Federal Regulations (CFR), Part 888, Sections 1100–5980, lists orthopedic devices together with their classifications (Class I, II, or III), which are defined in Part 860. Research in the field of orthopedics has resulted in an explosion of new materials and product concepts, many of which are not addressed under 21CFR §888. FDA has published guidance documents to address some of these new devices; however, in some cases, it is still necessary to contact the FDA directly for guidance on their classification. In the case of devices that blend modes of action or therapies, and for which a clear precedent has not yet been established, contacting the FDA Ombudsman for a ruling is required.

3.1. Class I Orthopedic Devices

Class I devices are not life-sustaining or life-supporting and, therefore, are subject only to general controls (i.e., device labeling, Good

Manufacturing Practices [GMPs], establishment registration, and device listing). All Class I devices require a 510(k) premarket notification unless individually exempted by the FDA, but do not normally require the submission of clinical data. An example of a Class I orthopedic device is a mixing device for polymethylmethacrylate cement used in cemented arthroplasty (888.4210). Cement mixing devices are exempted providing they are composed of the same materials as their pre-amendment counterparts; therefore, they do not require a 510(k) submission. However, AC-powered cast removal instruments (888.5960) are not exempted and a 510(k) submission is required.

3.2. Class II Orthopedic Devices

For Class II devices, general controls alone are not sufficient to provide reasonable assurance of their safety and effectiveness. Special controls, such as performance standards, postmarket surveillance, or other evaluative processes, are also required. Class II orthopedic devices range from arthroscopy systems (888.1100) to intramedullary rods (888.3020). Generally speaking, all Class II devices require a 510(k) submission. Although most 510(k)s do not require clinical data, if there are no preamendment counterparts or approved precedents or if the device will be used in a new application, clinical data may be required to support substantial equivalency claims. As an example, according to a recent guidance document, if an arthroscopy system is intended to be used in a different joint than previously approved for, clinical data will be required. On the other hand, the same arthroscope for use in a previously evaluated joint would not require clinical data to be included in a 510(k) submission. One should collect required clinical data for a 510(k) under an investigational device exemption (IDE), although, in some isolated cases, retrospective data and data from European clinical studies are deemed acceptable.

3.3. Class III Orthopedic Devices

Class III devices are significant risk devices that may be either life-sustaining or life-supporting, or for a use substantially important in preventing impairment of human health. These devices require reasonable evidence of safety and efficacy before they can be legally marketed. An IDE and a Premarket Apppoval Application (PMA) with clinical data must be submitted for all Class III devices unless exempted by a statement that no effective date has been established for the requirement of a premarket approval. In that case, those

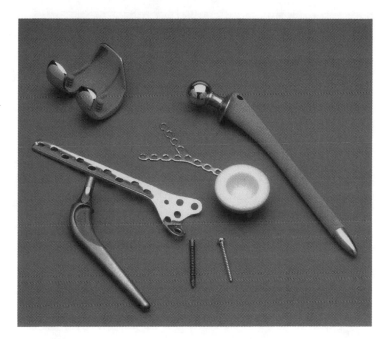

Fig. 2. Experimentation with different materials introduced various metals into orthopedic use, such as cobalt–chrome and titanium–aluminum–vanadium alloys.

devices may be submitted through a 510(k) premarket notification and, if found to be substantially equivalent to preamendment devices, can be legally marketed. The criteria for the submission of clinical data for these exempted Class III devices are the same as for Class II devices. Most noncemented, biologically fixed joint replacements are Class III orthopedic devices, except for porous-coated noncemented hip replacements, which were reclassified by petition to Class II.

4. Preclinical Device Evaluations

Before the 1976 Medical Device Amendments, orthopedic devices were essentially limited to joint replacement implants manufactured from metallic alloys. Experimentation with different materials introduced various metals, such as cobalt–chrome and titanium–aluminum–vanadium alloys (*see* Fig. 2), into orthopedic use. Although their use today is considered safe, as substantiated by years of safe clinical use, other materials did not fare as well. Teflon, a plastic evaluated early, failed catastrophically as an articulating surface in the hip joint.[2] Post-

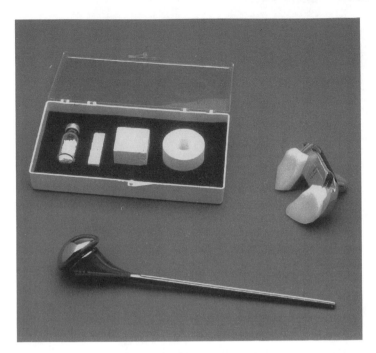

Fig. 3. Today orthopedics encompasses new alloys, composites, enhanced polymers, osteoconductive coatings, bioactive (osteoinductive) proteins, and a host of other esoteric concepts and materials.

amendment experimentation has also yielded some disastrous results involving such materials as polysulfone and carbon-reinforced poly-ethylene.[3, 4] This is a clear example of the fact that, even with the safeguards instituted through the 1976 Medical Device Amendments, there is a potential risk any time a new material is incorporated into clinical use. The importance of careful preclinical screening is only underscored as one performs a retrospective review of these clinical failures.

Because of the continuing safety and efficacy of the systems still in use today, the need for new materials in the field of orthopedics has not led to much experimentation except for the examples given above. Today, the field of orthopedics encompasses resorbable polymers, new alloys, composites, ceramics, enhanced polymers, osteoconduc-tive coatings, bioactive (osteoinductive) proteins, and a host of other more obscure materials (*see* Fig. 3). Clinical evaluations are frequently complex and must consider biocompatibility issues previously not addressed. It is impossible to understand the ramifications of the clinical evaluation of orthopedic devices without a brief discussion of the preclinical testing requirements.

The historical paucity of new materials in orthopedic device design has also resulted in the accumulated expertise of industry and the FDA in preclinical evaluation of orthopedic devices to be largely limited to efficacy screening (i.e., performance evaluations), or the use of the device in an animal model mimicking the actual intended practice. The FDA, through several guidance documents, clearly identified the types of preclinical testing requirements for a limited group of devices. Safety information is addressed in some instances and at times extrapolated from these models, but it is not comprehensively addressed, especially in situations where new materials are being considered.

Both the FDA and the orthopedic industry have struggled with the challenges of evaluating the safety of orthopedic device materials. Borrowing heavily from experiences with pharmaceutical screening, hosts of assays (some of them irrelevant and completely inappropriate) have been applied to the screening of new materials. The first attempt to identify and coordinate preclinical biocompatibility requirements for devices was released as the "Tripartite Biocompatibility Guidance," published by the FDA in 1986 to harmonize testing requirements in the United States, Canada, and the United Kingdom. This document incorporated guidelines published by the Health Industry Manufacturers Association (HIMA), American Society for Testing and Materials (ASTM), US Pharmacopoeia, the FDA, the American Dental Association, the British Standards Institute, and the European Confederation of Medical Suppliers Association.

Although this was the most comprehensive guidance of its kind, it still did not adequately address the inherent differences between drugs and devices, let alone the multitudinous disparities between different countries' requirements. In 1992, as an outgrowth of the harmonization process within the European Community, the International Standards Organization (ISO) released the most complete and all-encompassing list of requirements for device biocompatibility. The "Biological Evaluation of Medical Devices," ISO 10993 in 12 parts, is the most comprehensive standard to date that addresses the unique challenges of preclinical device testing and should be consulted as the definitive reference in evaluating preclinical testing needs. In fact, in response to the creation and adoption of this document, the FDA, as of July 1, 1995, replaced the Tripartite with this standard.

No guideline can be all encompassing, however, and there will be instances in which a device material will fall outside of defined parameters, either because of its composition or its intended use. This is another example of a situation where direct guidance on the part of FDA personnel must be sought. Inevitably, preclinical evaluations, no matter how sophisticated, can never fully reproduce or predict clinical experience.

5. Planning a Clinical Study

After determining that a device application will require the submission of clinical data and addressing preclinical testing needs, the next step is to carefully plan the clinical investigation. It is important to note that in many cases, FDA will approve an IDE prior to the completion of preclinical testing, providing that the critical toxicology screening proves negative and the remainder of the required biocompatibility battery is underway. All required preclinical evaluations, however, must be completed prior to submitting a PMA.

In designing a clinical study, it is important to be aware of the device's history. If the device to be evaluated has a precedent approved through the PMA process, then its Summary of Safety and Effectiveness (SSE) is a matter of public record, available according to the Freedom of Information Act. A review of a product's SSE provides information on the design of the study and the parameters measured. In the case of devices without a regulatory precedent, a review of the current literature can identify generally accepted methods of clinical assessment. In either case, a timely review of the literature is necessary because technology may have changed since the PMA approval. The FDA is admittedly not in the business of dictating clinical practice and any clinical evaluation that does not readily reflect standard practice in the conduct of medicine is destined for failure. One should also consult literature published by the FDA to reflect agency guidelines in the design of clinical trials.

The FDA's "Final Report of the Committee for Clinical Review" released in March 1993 (also known as the Temple Report), identified "...patterns of deficiency in the design, conduct, and analysis of clinical studies."[5] Dr. David Kessler, FDA Commissioner, remarked on their findings by highlighting a "...pattern of serious problems in study design—problems such as the lack of a clearly stated hypothesis, no clearly stated endpoints, and poorly chosen controls."[6] In addition, the Temple Report clearly described other problem areas in the design of studies, including deficiencies in the use of proper sample size calculations and bias-reducing techniques, such as blinding, randomization, and other scientific principles. The sponsor should carefully develop these clinical study elements, given the intense regulatory scrutiny they will likely receive.

5.1. A Clearly Stated Hypothesis

One can hardly consider the definition of a hypothesis for a clinical study without the definition of an endpoint or selection of adequate controls. For the purpose of discussion of the hypothesis alone, however, assume that both have been effectively defined. Virtually every orthopedic device or procedure has an existing precedent, and the vast majority of innovations merely consist of improvements on the existing technology. These changes are frequently minor and essentially undetectable during the prescribed 24-mo investigation, given the somewhat low sensitivity and the imprecise nature of the noninvasive tools for measuring success or failure. Thus, to claim statistically significant superiority is to guarantee rejection of the null hypothesis.

Successful orthopedic clinical studies have historically used the "as good as" null hypothesis. Showing significant positive improvements in a study with an "as good as" hypothesis is far more persuasive than showing no differences or, at best, slight differences with a "better than" hypothesis. It is important to note that the "as good as" null hypothesis is statistically indistinct from the "no worse than" hypothesis most frequently referred to by statisticians in the orthopedic research industry.

The "as good as" hypothesis touches on the issue of clinical utility. According to FDA "Memorandum on Clinical Utility," a clinical investigation in support of a PMA must show that the "...subject device has clinical utility; i.e., that the device has a beneficial therapeutic effect...."[7] Although historically the FDA has let market forces dictate pricing, recent attitudes expressed by agency personnel reflect a concern that differences of cost be proportionately reflected in outcome improvement without regard to the cost of supportive technology. The Health Care Finance Administration (HCFA) has further blurred the lines of financial jurisdiction by proposing that Medicare reimbursements be withheld for any and all investigational devices. These are important issues to keep in mind when designing a clinical investigation.

5.2. A Clearly Stated Endpoint

The endpoint in any clinical investigation has two components. One component relates to the duration of the study, but the other equally important element relates to whether certain parametric measurements, prequalified as indicators of success, have been attained. Historically, clinical trials of orthopedic devices have been arbitrarily required to last for a minimum of 24 mo. Yet it is naive to assume that a single endpoint can be applied equally to disparate devices and procedures. In practice, this length of time is probably either too long or too short.

If one is evaluating a device used in managing fractures, a 12-mo endpoint is generally more than adequate to assess its efficacy. However, a study designed to assess surface-coating differences between implants should logically employ an endpoint involving survivorship, a statistical phenomenon most accurately expressed in decades (not months). In short, an arbitrary chronological endpoint for all orthopedic medical devices is inappropriate. One must be prepared to present a logical argument to the FDA for a reasonable study endpoint, although the argument may be unsuccessful.

The second component of an endpoint is, as mentioned above, the tool used for measuring success or failure of a device or treatment. Surveys of available SSEs and published research yield valuable insights in this area. Hip and knee arthroplasty are evaluated both functionally and radiographically via several readily recognized and accepted schemes, all clearly documented and substantiated in the literature. Assessment of fracture healing, on the other hand, may not be as explicitly laid out in the literature. In cases in which measurement tools are not readily discernable, it is critical to ascertain standard clinical practice by obtaining consensus of practicing orthopedists. In pre-IDE meetings as well as in presentations to the FDA Orthopedic Panel, it is highly recommended to employ a credible authority in the particular orthopedic specialty to assist in validating the methodology proposed or used in a devices clinical evaluation.

The term *outcome* is the clinical research buzzword of the 1990s. It is used to describe patient satisfaction (quality of life), clinical success, or cost reduction but is, by definition, the measurement of a clearly defined objective in the final analysis. Most of the current published outcome-measuring tools primarily assess patient satisfaction, incorporating clinical outcomes as secondary parameters. John Ware, Ph.D., of the New England Medical Center in Boston developed the Short Form 36 (SF-36), which uses health status questions to assess a patient's pre and post treatment satisfaction.[8] The Health Status Questionnaire published by the Health Outcomes Institute also offers several disease-specific variations. Many ambitious clinical scientists attempt to incorporate outcome evaluations into clinical trials. The controversy surrounding the clinical utility of patient satisfaction may encourage companies to amass patients' psychosocial data to substantiate their claims of successful treatment. These evaluations may include everything from Visual Analog Scale (VAS) determinations of patients' "quality of life" to their ability to walk their pets or play golf.

At the risk of undermining the importance of this new and interesting area of clinical research, it is critical to emphasize that measurement of outcomes have applications in clinical trials *only* when used with already validated parametric tools. Using them as definitive determiners of success or failure only increases the risk of FDA rejection of the sponsor's application.

In light of this recent focus on outcomes research, it is important to review the difference between *efficacy* and *effectiveness*. Efficacy is a measure of "how well a particular medical intervention works within a very controlled clinical trial"; effectiveness, on the other hand, is a measure of "how does the intervention work when everything isn't perfect."[9] Efficacy relates to a homogenous patient population, specifically identified parametric tools, and all the other research elements discussed in this chapter. Effectiveness relates to the more esoteric issues of patient satisfaction, clinical utility, and other elusive issues that outcome measurements attempt to quantify.

5.3. Properly Chosen Controls

Orthopedic device studies, by nature of their delivery systems, preclude the possibility of using placebo controls. However, because many orthopedic devices treat maladies for which alternative therapies are available, selecting a control group is generally straightforward.

The most appropriate control is the current accepted standard of care. Examples of control treatments are replacement of the central third of the patellar tendon for studies on anterior cruciate ligament replacement, autogenous bone in the case of bone void filler studies (bone-graft substitutes), and cemented implants when the cementless counterpart is being investigated.

In contrast, what is the appropriate course in the case of a clinical diagnosis for which the existing treatment is undesirable? For example, what does one use as a valid control for a cementless-ankle joint replacement? The orthopedic community is divided on the clinical utility of currently available cemented-ankle replacement systems, with most expressing skepticism regarding the fate of any ankle replacement system. The only alternative treatment for chronic ankle arthropathy is fusion of the joint. Thus, the most appropriate comparison of treatments for this joint would be between an ankle-replacement system and ankle fusion.

Although the FDA is categorically opposed to studies that do not employ a concurrent control, several arguments surface for using historical (literature) controls. Many scientists maintain that these controls are relevant and adequate. They argue that, given the rapid developments in orthopedic research, subjecting a cohort of patients to a clinically "obsolete" standard of care is denying them the latest available technology. Because new techniques and devices may take almost a decade to be approved, there is some validity to this argument.

Another important point to acknowledge is that market forces influence even this critical component of clinical study design. When the alternative treatment is manufactured by a competitor, the sponsor is unlikely to propose that treatment as the control for a clinical investigation. In those and other rare cases, historical literature controls have been permitted, albeit reluctantly.

Collecting remuneration for investigational devices is another issue that bears discussion in light of establishing appropriate controls. In many ways, this issue is unique to the orthopedic industry. Given the fact that in the preamble to the IDE regulations a manufacturer is discouraged from using the IDE as a subterfuge for unapproved commercialization of a device, along with the clear precedent of not collecting funds in pharmaceutical clinical trials, many health institutions and clinical investigators (including HCFA) have strongly objected to fees being charged for investigational orthopedic devices. However, several very important differences exist between the intrinsic nature of the orthopedic device and its cousin, the investigational drug. Industry reasoning for this policy is valid, as follows:

1. Orthopedic research frequently involves improvements or innovations based on the current standard of care (approved devices). Although safety is always a concern, the focus is mainly on efficacy. Thus, patients are placed at less risk than patients evaluating the safety of a new drug; yet they receive the benefit of new and otherwise unavailable technology.
2. In most cases, the device itself (unlike the drug per dose) represents a substantial intrinsic cost on a per unit basis. Although pharmaceutical companies make comparable investments into research, they recoup their investments within months of drug approval. In contrast, return on investment for orthopedic devices is measured in years.
3. Congress, as well as the FDA, (but unfortunately *not* HCFA) realizes the disparity between the individual per unit costs represented in the manufacturing of medical devices vs drugs. Although IDE regulations include provisions for recouping the costs of the device within the context of the clinical study, they also include a proviso against realizing a profit during the IDE phase.
4. Given the above *status quo*, not collecting remuneration for an investigational device could be construed as enticement, a practice strictly guarded against in the IDE regulations.

Although the IDE regulations do not specify that the investigational device must not be priced higher than the control, it should be obvious that this is common practice. Failure to normalize the costs across cohorts will undoubtedly bring strong resistance from the responsible institutional review board (IRB).

One must also note that the practice of collecting remuneration for orthopedic investigational devices is limited to instances where the device represents an improvement on a commercially available system. Technological and biological innovations have yielded products, such as artificial (or allografted) ligament replacements, bone-graft substitutes, bone-growth stimulators, and other treatments, for which commercially available alternatives do not exist. In these cases, attempting to recoup the costs of manufacturing the device within the confines of clinical study is not an option.

5.4. Adequate Sample Size Calculations

The "Medical Device Clinical Study Guidance" released by the FDA in 1993 clearly describes and defines formulas for calculating sample size for clinical studies.[10] Although individuals have different philosophies for applying these equations—specifically when dealing with the issues of binomial comparisons vs parametric comparisons—certain basic assumptions are valid for all sample size determinations.

Most orthopedic studies are unique because of the nature of their populations. Joint replacement studies involve older populations for which actuarial issues must be planned into the sample size calculations. Generally, expectations during a period of 2 yr are that 5% of the population will be lost because of death and 15% will be lost to follow-up for other reasons. On the other hand, although trauma studies involve a younger population base, they frequently include a high percentage of transients. This factor adversely affects anticipated compliance and must be planned for when calculating the total study population.

A large sample size is generally undesirable because of the higher cost and the additional time required for the completion of the study. Two factors that most critically affect patient population size are definition of the null hypothesis and the statistical power used (the probability that the null hypothesis will be rejected if it is truly false). Both of these factors interrelate to the point where it is impossible to evaluate one without considering the other. For example, if a null hypothesis is ambitiously stated as "better than," the statistical power will necessarily need to be lower and the sample size higher. Larger sample size is necessary because sensitivity of the observations used to determine success/failure is generally very low, and it will take both a less-restrictive power as well as a large group of cases to show measurable differences. This is probably the most compelling reason for using the "as good as" hypothesis; it needs fewer cases to prove itself, necessarily employing a less restrictive power.

A "given" in the process of defining sample size for a clinical study is the homogeneity of the target population. In general, orthopedists do not wish to limit their population through stringent selection criteria because of the evident implications of such a decision. If a variety of diagnostic categories are included in the population, stratification by these diagnoses will undoubtedly be required for data analysis. Not only is it possible, but it is very likely that certain groups may have inadequate numbers for statistical comparison and that the indication for that diagnosis may not be approved. Limiting the allowable diagnoses can also be used to positively affect the outcome by eliminating groups at risk. On the other hand, FDA approval will only apply for those groups that have been evaluated clinically, and limiting the population in the study may also limit the potential market for the device.

5.5. Bias-Reducing Techniques

The three most significant contributors to reducing bias within a clinical investigation include the use of blinding (also referred to as masking), randomization, and employing third party evaluators. Proper patient selection, standardizing parametric observation methods, and ensuring equal compliance with the clinical plan across all clinical sites can also have a profound effect on reducing or eliminating bias.

5.5.1. Blinding

The implausibility of using a placebo in an orthopedic device study for blinding purposes has already been discussed, but in almost all cases blinding the patient to treatment is possible, and this should be boilerplate in all clinical designs. Notable exceptions occur when comparing noninvasive to invasive therapies or when comparing treatments in which one may require additional surgical intervention. In these cases, the patient is patently aware of the additional surgical site resulting from the treatment.

Blinding the investigator can prove more challenging. Impairing a clinician from visualizing an implantable device would have obvious and disastrous surgical consequences. Secondary blinding, however, is possible during the follow-up process and can be achieved by employing a third party for postoperative evaluation. Tertiary blinding can also be used to blind the investigator to the implanted device by having a separate third party conduct radiographic evaluations.

5.5.2. Randomization

Randomization is key to eliminating bias. In certain situations, however, convincing investigators of its necessity might be difficult, especially if they have argued against the concept of concurrent controls. Orthopedic surgeons, by practice, specifically select treatments for their patients based on their previous personal experiences with outcomes in selected populations. A close monitoring of randomization procedures to ensure compliance is mandatory in a good clinical study.

5.5.3. Third-Party Evaluators

The use of third-party evaluators is central to the concept of blinding and has notable utility when conducting diagnostic assessments based on laboratory or imaging data. Not only do they eliminate bias on the part of the investigator, they virtually eradicate intercenter variability in interpreting the results. This is true only where differences between the clinical sites in analyses, imaging quality, and reporting are not issues. Using a third-party evaluator can also provide data to compare with individual investigator interpretations as a quality control check for intercenter variability.

5.5.4. Patient Selection

Selection of the patient population based on clinical criteria for sample homogeneity is central to the microdemographic component of sample size determination and can also have a profound effect on reducing bias. However, the macrodemographic effect from patient differences found between geographic regions, both domestically as well as abroad, must not be underestimated. If patients in an investigation are limited to discrete geographic locations, the patient pool will not reflect all potential populations for which the orthopedic device is intended. A proper mix of urban and rural cases incorporating the differences found across topographical and climatic regions can help ensure homogeneity on the macrodemographic scale.

For reasons unique to the orthopedic industry and for other financial concerns covered previously in this chapter in Section 5.3., financial inducements are generally not employed in the recruitment of patients for orthopedic clinical investigations. Exceptions do occur, mainly in the field of fracture management studies, where the transient nature of the population necessitates an incentive for compliance.

5.5.5. Selecting Parametric Observation Methods

The importance of parametric observation selection to the determination of study endpoints is obvious, but it can also have an impact on bias. The use of standard recognized tools is imperative to reduce variability between investigational sites. Clinical observations in orthopedic studies can encompass functional parameters, laboratory or radiographic evaluations, and the reporting of complications. Many of the orthopedic subspecialties have developed evaluative tools to standardize postoperative clinical observations. These become standard practice through dissemination in peer-review journals and other publications.

Most functional evaluation schemes cover three basic areas: assessment of pain, range-of-motion (or measurable clinical function), and activities of daily living. For evaluating total hip replacement surgery, the Harris Hip Scoring System, the Hospital for Special Surgery Scoring System, and the Société Internationale de Chirurgie Orthopédique et de Traumatologie (SICOT) Scoring System are examples of recognized tools currently in use. [11] Failure to obtain input from prospective investigators to determine which system is most commonly used may result in noncompliance with collection of clinical data. In designing case report forms (CRFs), modifying the accepted evaluation systems to accommodate variations in practice or concentrating on areas specific to features of the orthopedic device being studied is acceptable.

Radiographic assessment allows less consensus and is prone to greater subjectivity than other evaluation systems. Certain parameters, such as radiolucencies and osteolysis, are central to all radiographic observations. However, other than specific references by Gruen[12] describing zones for radiographic observations, and Brooker[13] identifying degrees of heterotopic ossification, the literature is somewhat lacking in standardized evaluative tools. In these cases a preinvestigation meeting is crucial to determine which parameters are indispensable to the participating investigators. To minimize observational variations, employing an independent radiologist to read all films is suggested.

5.5.6. Ensuring Compliance

Finally, equilateral enforcement of compliance requirements is one of the most underestimated influencers of bias. Tolerance of noncompliance at selected investigation sites will result in data deemed ineligible for pooling. Ways to enhance compliance are discussed in greater detail in Section 6.4.

6. Managing a Clinical Study

Realizing that conducting a clinical study occurs at the investigational site and that the sponsor merely manages this process is very important. Management must never be interpreted as mere oversight; micromanagement in this instance is not only permitted, but required.

6.1. Selection of Investigators

Before discussing the process of selecting investigators, the definition of a clinical investigator's role in the clinical study must be addressed. Managing a clinical investigation in orthopedics does not begin with IDE approval; it should be well underway by that time. It actually begins well in advance of IDE submission, at which time the potential clinical investigators should have been selected, identified, and screened.

To fully understand the importance of this requirement, another of the many significant differences between clinical studies for drugs and devices must be highlighted. The administration of a pharmaceutical agent is detailed in a clinical plan that has been finalized prior to investigator selection. The role of the drug study investigator is complete as soon as he or she writes the prescription. For device studies, the administration or delivery of an orthopedic device constitutes the most critical step of an investigator's involvement. Thus, selection of the investigator is the most important factor that can contribute to the success or failure of the clinical investigation for a device.

Implantation of orthopedic devices involves specialized orthopedic surgeons performing delicate procedures. The clinical success of an orthopedic device is highly dependent on the technical skill of the surgeon. An inferior device implanted by a superior technician is most often more effective than a superior device placed by an inferior technician. Variability in the technical ability of an investigator to effectively deliver the orthopedic device is undesirable and can have devastating effects on a clinical investigation. Therefore, the selection of orthopedists who specialize only in the technique under investigation is highly recommended. Not only does this help to eliminate intercenter variability and potential complications, it can also reduce the learning curve phenomenon.

It is critical to involve the clinician early in the process of study design. This involvement not only enhances compliance by giving the participant ownership in the process, but also educates the investigator to both the investigational requirements and to the importance of adhering to each requirement. One of the most significant and necessary contributions each investigator makes to the clinical study design is selecting parametric measurements for patient assessment. The sponsor is heavily dependent on the input of the investigator to identify current medical practice as it applies to the delivery of the device in question. If the design of a study does not closely parallel the current standard of care, it is not likely that study compliance or a healthy accrual rate will be achieved or maintained, and it could be argued that the patient population is being deprived of adequate medical care. The most effective and efficient way to accomplish this is by holding a preinvestigation meeting with all proposed investigators. In this forum, the sponsor can review the basic outline of the protocol and attempt to build consensus regarding the details of study design.

Now that the importance of the investigator to the design and successful conduct of the clinical study has been established, the process of recruitment of clinical investigators should be discussed. The role of marketing in the selection of investigators is one that requires a *maximum* of restraint. Recruiting clinical investigators for ulterior marketing motives invariably spells disaster. Although clinical research departments rely heavily on the input of marketing to identify clinicians with the specific skills needed for a clinical study, the selection process usually works best when the initial contact and screening is handled solely by the clinical research group.

Once a viable investigator has been identified, he or she may have additional recommendations for the role of the investigator. Another means of identifying potential investigators is to review the literature for clinicians experienced in the required procedures. Another effective mechanism is to liaise with market research personnel to determine which institutions have the largest volume of procedures. Many institutions now also have patient registries from which diagnostic demographic data can be extracted and provided.

6.2. IRB Approval and Informed Patient Consent (IPC) Issues

Institutional Review Boards, 21 CFR Part 56, outlines specific requirements for the membership, responsibility, and authority of the IRB. Unfortunately, the ratio of FDA compliance personnel to IRB does not allow for comprehensive surveillance. Limited random audits have historically revealed deficiencies in more than 25% of those inspected.

The FDA, in apparent anticipation of this problem, places responsibility for IRB compliance on the sponsor by including the requirement for IRB compliance in the IDE regulation. This creates an awkward situation, given the strict limitation on interaction between the sponsor and the IRB as is frequently imposed on the sponsor by the IRB to avoid any semblance of collusion.

An effective mechanism for placing this responsibility on the IRB is to merge a boilerplate statement certifying compliance with 21 CFR Part 56 with a certification form for study approval and reapproval. This form, although not required by the IDE regulations, references 21 CFR Part 56 and states that the reviewing IRB is in compliance with it. At the same time, it certifies that the IRB has reviewed the clinical study and has approved or reapproved it. Signature by the IRB chairperson would certify both of these facts, and a copy of the form retained in the sponsor's study file would document compliance with the IDE regulations.

The above mechanism may soon be moot in light of the increased level of sophistication seen in IRBs in recent years. Many IRBs have now begun to fully exercise their responsibility and authority by dictating composition of and demanding final draft approval of the IPC, and asking for detailed reviews of the study before approving—and on an ongoing basis before reapproving—the investigation. It is not unheard of for an IRB to request a representative of the sponsor to present a clinical study summary during a convened session and/or to be available to answer questions. It is important to note that the presence of a sponsor representative during voting on approval is never permitted.

The portion of the investigational plan that receives the most scrutiny from the IRB is the IPC. Clinical research in orthopedics does not place the patient at as great a risk as other investigational procedures. At the same time, the patient does receive benefit of new and otherwise unavailable technology. Unfortunately, the patient may be unduly alarmed at the information that must be presented in the IPC, especially since it must also describe in detail the risks associated with treatment, regardless of the device. Most patients presenting for routine joint arthroplasty would never be confronted with as comprehensive a list of potential complications from hip surgery as those required by law in the IPC. Furthermore, many of the "new breed" of IRBs are requiring indemnification statements in an effort to limit institutional liability. Together, this can result in a document that can easily overwhelm and dissuade a potential subject from participating in a clinical trial. It is critical to review consenting procedures with investigators and their staff and sensitize them to this issue to prevent this from happening.

6.3. Data Collection and Monitoring

Frequent and meaningful communication with each clinical site is necessary to facilitate data collection as well as to assess the level of compliance and to be aware of problems in a timely manner. Documenting each contact with a memo to file or a telephone log is an effective mechanism for accomplishing this. A recent report of audits conducted by the Bioresearch Monitoring group at the FDA revealed a lack of documentation of site contacts as one of the most frequently cited deficiencies.

Another area of concern is the timely submission of clinical data. Delinquency in completion of CRFs can easily lead to retrospective collection of clinical observations, a practice that—in addition to being a flagrant protocol violation—can adversely affect the validity of the data being measured. A good rule of thumb is the "10 working day" requirement. This period, which roughly translates into 2 wk, should easily encompass any delays in obtaining missing information, mailing time, or other problems.

In addition, data review and screening will undoubtedly reveal incomplete or inconsistent data fields. The data monitor must refrain at all times from making assumptions in cases of missing data. Methods of dealing with data edits range from source document archiving with on-site copy initialing to telephone corrections. The mode of processing data edits is less important than the consistency with which it is done. On rare occasions, instances will arise in which assumptions regarding individual data fields must be made. The cardinal rule here is as follows: If the data are baseline (or preoperative), the assumption must be the most favorable option; if the data are postoperative, the assumption must be the least favorable. This way, if biasing should occur, it will always be against the device being evaluated.

To ensure compliance with the investigational plan, the IDE regulations suggest annual visits to each investigational site. Experience, however, dictates conducting more frequent visits, especially during the active enrollment process. Most protocol violations occur during this phase of the study, and the ability to correct and anticipate these problems is not exercisable through hindsight. Each visit should culminate with a closeout, face-to-face meeting with the investigator.

The availability of the investigator to meet with the monitor should be the limiting reagent in scheduling a site visit.

6.4. Compliance Issues

Compliance within the scope of the clinical investigation involves adherence to the clinical protocol and can be divided into investigator compliance and patient compliance. The most frequently cited investigator compliance issues are failure to properly obtain IPC, improper management of investigational inventory, and initiating a study prior to IRB approval. Patient compliance issues deal exclusively with their adherence to the required schedule of follow-up visits.

The most frequent problem with IPCs is failure of the patient, investigator, or witness to date their signature on the document after the date of surgery. An important distinction to be made at this point is the difference between the two elements that make up the consenting process. The IPC is merely a document attesting that the process of informing and obtaining the consent of the patient was carried out. The document is of secondary importance to the process itself. The process involves an active dialog in which all of the patients' concerns regarding their involvement in the study and the procedures to be performed are supposed to be addressed. In cases where deficiencies with the execution of the actual document exist, it is frequently possible to extract from clinical or hospital notes the fact that this discussion did indeed transpire and thus substantiate the process. The FDA's main concern with IPC issues—and consequently an area of great scrutiny—involves the fear that a patient might unknowingly receive an investigational device. In actuality, this is extremely unlikely but, should it occur, the incident must be reported to FDA and the reviewing IRB within 5 d of occurrence.

Management of the investigational inventory necessarily involves personnel who, in many cases, are not under the jurisdiction of the investigator. Careful training of these individuals is important, as well as emphasizing to the investigator that, regardless of his or her degree of control over the situation, it remains a clearly documented responsibility.

Initiating patient accrual at a clinical site prior to IRB approval in today's world of orthopedic research is inexcusable and represents the height of negligence, both on the part of the sponsor and the investigator. It is most likely a portent of even more severe compliance issues. Even though its occurrence is increasingly rare, FDA has been historically sensitized to it. Agency concerns can be alleviated by incorporating the names and addresses of all reviewing IRBs and the date of study approval by the IRB, together with the date of the first implant into the required 6-mo current investigator list [21CFR §812.150(b)(4)].

Patient compliance, on the other hand, is an area that involves maximum cooperation and coordination among sponsor, investigator, and study subject. Because most orthopedic devices are long-term (hopefully, permanent) implants, follow-up must be planned accordingly. One should thoroughly cover this matter with the investigators at the preinvestigation meeting to ensure that clinical follow-ups mimic actual practice. This is especially important because, in most cases, the sponsor does not cover any costs of patient treatment, and any variation from usual and customary practice could appear suspect to third-party payers. As mentioned before, the expected follow-up for orthopedic implants is 24 mo. With most orthopedic devices, this consists of a preoperative evaluation, radiographs in the immediate postoperative period, and follow-ups at 3, 6, 12, and 24 mo.

The FDA considers a patient lost within an interval if the patient reaches the interval anniversary date and has not returned for a follow-up. If a patient returns after the anniversary date, however, he or she may still be counted for that interval at that time. It behooves the sponsor to have patients return as close to their anniversary dates as possible and preferably before them. The FDA requires that an 80% compliance level be maintained at each follow-up interval. At least 85% of 2-yr data is required for a PMA submission, and all cases must have reached their 2-yr follow-up before a final report can be submitted.

Both investigator and patient compliance issues remain clearly the responsibility of the investigator. This is something that must be reinforced during the selection and training of each investigator. Perhaps the single most important truth to keep in mind when evaluating compliance issues is that *no investigator is indispensable.* Having the wisdom and the will to replace noncompliant or ineffective investigators is perhaps the most powerful enhancer of overall compliance in a clinical investigation.

6.5. Reporting of Medical Events

The monitoring of complications in orthopedic device trials can be very challenging. Many variables are introduced through the actual process of administering (implanting) the device. Because the follow-up time is so long, experience dictates the "when in doubt, report" rule; that is, to collect and report all medical events, no matter how trivial or irrelevant they may seem, and sort them out in the final analysis. FDA reviewers will be assured that no medical events were ignored and will help to defuse the issues of unreported complications.

For classification purposes, medical events are generally divided into operative and postoperative complications. They are reported as local (operative site) and general (systemic) complications. In the collection and reporting of medical events, it is important to include the date of onset, course of therapy, and the patient's current status to determine whether the condition has resolved. Adverse device effects (ADEs) occur in situations in which the patient experiences a health-threatening failure (structural compromise) of the device itself. One must report all ADEs to the FDA within 10 working days.

7. Case Study: The New Jersey LCS® Total Knee System with Porocoat Coating

No single orthopedic device can, as a case study, illustrate all the possible esoteric scenarios that can be encountered when planning and conducting a clinical study. The regulatory history of the New Jersey LCS Total Knee System with Porocoat Coating (now known as either the New Jersey LCS Total Knee System or LCS Knee) is one example of the decisions that must be addressed. The LCS Knee, like many orthopedic devices, has expanded from a single design to a product system that offers options for a variety of arthritic conditions. What differentiates it from the host of other prosthetic devices is that the additional options are simply configurations, not replacements of the original design. In fact, it is the only FDA-cleared mobile-bearing knee replacement, as well as the only uncemented knee replacement still in use in its originally cleared design. The regulatory challenges of obtaining approval for not only the original design, but for the additional iterations, provide a practical application of the IDE regulations and FDA guidelines for conducting a clinical study.

The original IDE for the LCS Knee was for cemented use and was filed in 1980; it subsequently went through the PMA process and was approved (in two separate device configurations) in 1985. Cementless application for additional configurations were approved as supplements to the original cemented PMA (1990, 1994), because the FDA deemed that they were essentially the same device, differing only in the modes of fixation. The success of each of the composite clinical investigations resulted in FDA approval being readily obtained, and the attainment of the largest series of knee cases with the longest follow-up in the history of orthopedic research.

Following is a review of how the various segments of the LCS Knee submissions dealt with each of the requirements for a successful investigation, illustrating how the increase in the level of sophistication of research over time has affected the quality of the clinical data collected.

7.1. A Clearly Stated Hypothesis

For the original cemented IDE, the hypothesis was that the system would be "as good as" other systems currently in use. The hypothesis was essentially the same for each of the subsequent cementless IDEs.

7.2. A Clearly Stated Endpoint

The hypothesis for both the cemented and cementless trials was to be evaluated by standard parametric observation methods at an endpoint of 24 mo, the customary FDA-imposed duration for all orthopedic clinical trials. Again, this practice has not changed over time.

7.3. Properly Chosen Controls

The original cemented IDE illustrates one of the few instances in which literature controls would be considered. Since the LCS Knee is a mobile bearing knee replacement and had no equivalent approved counterpart for comparison at that time, there was no appropriate prospective control available. Instead, literature reports of other knee systems were used for statistical comparison. When it came time to submit a PMA for cementless configurations, the cemented cases from the previous IDE/PMA were used as retrospective controls for the functional data; literature controls were used for the radiographic comparison.

7.4. Adequate Sample Size Calculations

When the original IDE was submitted for the cemented LCS Knee, there was no requirement for a sample size calculation. The emphasis of the FDA at this time was on limiting accrual of cases to minimize the risk to the population. For the cementless IDE, a clearly defined control was available (albeit retrospective) in the cemented population, and a requirement for a minimum of 100 cases at 24 mo per device configuration was defined in the IDE (a total of over 800 cases were accumulated in the study). For the most recent IDE on the LCS Knee (filed in 1991), a sample size calculation based on the formula provided in the FDA guidance document (and based on an estimation of outcome from the previous PMA populations) yielded a sample size of 168 cases/treatment (test and control), or a total of 336 cases.

7.5. Bias-Reducing Techniques

7.5.1. Blinding

The early cemented IDE did not involve a prospective control, eliminating the possibility for blinding. The first cementless IDE also used historical controls, precluding the possibility of blinding. The most recent IDE comparing surface coatings employed prospective controls and was able to blind not only the patient to treatment but, through a specialized handling technique, was also able to mask the surgeon regarding patient treatment.

7.5.2. Randomization

For the reasons described above, the early IDEs could not employ this scheme. The most recent IDE was able to incorporate a randomization scheme into the plan for patient treatment.

7.5.3. Third-Party Evaluators

The use of a third party evaluator would have been possible with the early IDEs in the review of clinical radiographs, but was not identified as a bias-reducing alternative until the late 1980s. The latest IDE involved a clearly defined radiographic protocol and employed an independent radiologist blinded to treatment and center.

7.5.4. Patient Selection

The earlier IDEs did not collect sufficient rheumatoid or revision cases to warrant approval for these diagnostic subsets. Subsequent IDEs limited the diagnoses to cases with noninflammatory degenerative joint disease (or osteoarthritis) presenting for primary surgery. Both the cemented and cementless IDEs involved multiple sites (nearly 20) from a variety of geographical locations, but the most recent IDE limited patient accrual to six clinical sites. Efforts were made to scatter these representatively across the United States.

7.5.5. Selecting Parametric Observation Methods

The first IDEs used a functional and radiographic observation scheme designed by the developer of the LCS Knee. The most recent

IDE judiciously employed the Knee Society rating system, one that had already been validated through numerous published reports.

7.5.6. Ensuring Compliance

Compliance for PMA purposes was previously considered adequate at 80%; current FDA guidelines require 85%. Adequate compliance was maintained throughout all phases of the studies through diligent investigator selection and management, as well as frequent communication and monitoring.

8. Recommended Reading

Finding references that deal strictly with clinical investigations in orthopedics can be somewhat challenging. As with the clinical investigation of any device, one must draw the basis directly from the IDE regulations. This should be the template for identifying all clinical study design, conduct, and reporting requirements. The "Investigation Device Exemption Manual" (92-4159) is a required desktop reference that effectively summarizes the responsibilities of the sponsor. This manual is published by the FDA and is available through the Division of Small Manufacturers Assistance (DSMA). Further information can be found in other FDA literature. The DSMA provides a listing of all available FDA publications and will supply most publications at no cost. "Medical Device & Diagnostic Industry," a trade journal published monthly by Canon Communications, Inc., Santa Monica, CA, is also available free of charge to most qualified subscribers. This journal routinely reports regulatory issues pertaining to medical devices. For more clinically focused information, the *Journal of Bone and Joint Surgery* and *Clinical Orthopedics and Related Research* are peer-reviewed journals whose copies, which feature orthopedic clinical research, are available through any medical library.

9. Summary

The preceding information in this chapter was intended to provide a perspective on the application of general IDE requirements to orthopedic research. Although not all issues have been completely addressed, the most important factors for a successful endeavor in orthopedic research have been covered.

The first and most important factor is proper study design. No amount of oversight and management can overcome a poorly planned investigation. It will affect everything from accrual rate through compliance to the analysis of data. Special attention to the issues covered herein will not only ensure a good investigational plan but will also address FDA sensitization and help expedite the IDE-approval process.

The selection of investigators through identification, recruitment, interaction with, and training is another factor that determines good research. The selection process is the beginning of managing a successful clinical study. A clear identification of the requirements carried through into implementation of the clinical plan is central to staying on top of an investigation. Judicious management of each investigator, including termination and replacement if necessary, is the most powerful enhancer of compliance.

The final factor, although not specifically stated, has been alluded to throughout; it is an open line of communication with the FDA. Although frequent and frivolous inquiries are not in any company's best interest, responsible and reasonable dialog builds trust, increases comfort level, and establishes relationships that can define a company's image. Openness and disclosure on the part of the sponsor not only earn a reputation for honesty and integrity, but they also provide the FDA with input that can help them better address the unique research concerns of the orthopedic industry.

Additional Reading

Bradham, D. 1994. Outcomes research in orthopaedics: history, perspectives, concepts, and future. *J. Arthroscopic Rel. Surg.* 10:493–501.

Feldman, M. 1993. Issues in the design of clinical trials. *Med. Dev. Diag. Ind.* Nov.:96–102.

Food and Drug Administration. 1995. *General Program Memorandum Re: Use of International Standard ISO-10993*. May 1, Rockville, MD.

Huiskes, R. 1993. Failed innovation in total hip replacement. Diagnosis and proposals for a cure. *Acta Orthop. Scand.* 64:699–716.

Krischer, J., Hurley, C., Pillalamarri, M., Pant, S., Bleichfeld, C., Opel, M., and Shuster, J. J. 1991. An automated patient registration system and treatment randomization system for multicenter clinical trials. *Control. Clin. Trials* 12:367–377.

Johnson, L. 1994. Outcomes analysis in spinal research. How clinical research differs from outcomes analysis. *Orthop. Clin. North Am.* 25: 205–213.

Kahan, J. S. 1991. FDA restrictions on the commercialization of investigational devices. *Med. Dev. Diag. Ind.* June:80–83.

Lamirand, R. *One-Hundred Years of Excellence.* Depuy, Inc., Warsaw, IN.

Melkerson, M., Yahiro, M., and Mishra, N. 1993. Regulatory perspective for orthopaedic devices. In *Biological, Material, and Mechanical Considerations of Joint Replacement* (B. F. Morrey ed.). Raven, New York.

Nowak, R. 1994. Clinical trial monitoring: hit or miss? *Science* 264:1534–1537.

Nowak, R. 1994. Problems in clinical trials go far beyond misconduct. *Science* 264:1538–1541.

Stark, N. 1991. How to organize a biocompatibility testing program: a case study. *Med. Dev. Diag. Ind.* June:68–75.

Van Vleet, J. and Sherman, M. 1991. The history of institutional review boards. *Reg. Aff.* 3:615–628.

References

1. Waugh, W. 1990. *John Charnley: The Man and the Hip.* Springer-Verlag, London.

2. Charnley, J. 1979. *Low Friction Arthroplasty of the Hip.* Springer-Verlag, London.

3. Trentacosta, J. and Cheban, J. 1995. Lipid sensitivity of polyaryl-ether-ketones and polysulfone. Presented at the 41st Annual Meeting, Orthopaedic Research Society, February 13–16, Orlando, FL.

4. Wright, T. M., Astion, D. J., Bansal, M., Rimnac, C. M., Green, T., Insall, J. N., and Robinson, R. P. 1988. Failure of carbon fiber-reinforced polyethylene total knee replacement components. A report of two cases. *J. Bone Joint Surg. Am.* 70:926–932.

5. Food and Drug Administration. 1993. *Final Report of the Committee for Clinical Review. "The Temple Report."* FDA Report. March, pp. 1–45.

6. Kessler, D. 1994. FDA's revitalization of medical device review and regulation. *Biomed. Instrum. Technol.* 28:220–226.

7. Food and Drug Administration. 1991. *Memorandum on Clinical Utility.* Rockville, MD.

8. Ware, J. and Sherbourne, C. 1992. The MOS 36-Item Short-Form Health Survey (SF-36). I. Conceptual framework and item selection. *Med. Care* 30:473–483.

9. Gunter, M. 1995. "Past lessons: future promises." Keynote Address, Velocity User Meeting and Outcomes Conference, October, Minneapolis, MN.

10. Food and Drug Administration. 1993. *Medical Device Clinical Study Guidance.* Rockville, MD.
11. Johnston, R. C., Fitzgerald, R. H., Harris, W. H., Poss, R., Muller, M. E., and Sledge, C. B. 1990. Clinical and radiographic evaluation of total hip replacement. A standard system of terminology for reporting results. *J. Bone Joint Surg. Am.* 72:161–168.
12. Gruen, T., McNeice, G., and Amstutz, H. 1979. "Modes of failure" of cemented stem-type femoral components. A radiographic analysis of loosening. *Clin. Orthop.* 141:17–27.
13. Brooker, A., Bowerman, J. W., Robinson, R. A., and Riley, L. H., Jr. 1973. Ectopic ossification following total hip replacement. Incidence and a method of classification. *J. Bone Joint Surg.* 55-A:1629–1632.

7

Long-Term Evaluation
of Total Hip Arthroplasty

Frederick J. Dorey

1. Introduction

Total hip arthroplasty is one of the most successful medical procedures available today. In the short term, the patient receiving an artificial hip implant can almost always be guaranteed substantial pain relief, greatly improved range of motion, and an improved quality of life. Unfortunately, many patients will, over time, experience a subsequent loosening of their prosthesis and a return of the clinical symptoms that led to the original arthroplasty. The failure of the original (or index) surgery becomes complete when additional surgery is required, usually leading to a revision surgery (the removal of the original implant and insertion of a new prosthesis). Generally, the results of such a revision surgery are not expected to be as good as with the original surgery. However, most patients will not require such a revision surgery until many years after the index surgery, and many older patients will live their lifetime without the need for any subsequent surgery.

Thus, evaluation of the effectiveness of total hip arthroplasty as a medical procedure requires following patients for an extended period of time, usually 10 yr or more. The long-term evaluation involves the usual issues that are common to the evaluation of all medical devices and statistical analysis (i.e., adequate description of the patient population, evaluation of prognostic factors, stratified or multivariate

From: *Clinical Evaluation of Medical Devices: Principles and Case Studies*
Edited by K. B. Witkin Humana Press Inc., Totowa, NJ

analysis, use of confidence intervals as well as hypothesis tests, and so forth). However, the need to follow patients for an extended period of time as well as the elective nature of this surgical procedure introduces additional problems and issues.

Many of the issues involved in the long-term evaluation of medical devices can be illustrated by an evaluation of some of the more common cemented stemmed arthroplasties used in the 1970s and 1980s. The clinical results of these prostheses have recently been reported in the orthopedic literature.[1] Many important issues are omitted in the medical literature, such as proper study design, effectiveness criteria and outcome measures, methods of analysis, and treatment of patients lost to death or follow-up. For a properly conducted long-term evaluation of a medical device, however, it is important to address these issues.

2. Issues in the Analysis of Total Hip Arthroplasty

Prior to conducting an evaluation of any medical device usually several questions must be answered.

2.1. What Information Exists in the Present Literature?

No statistical analysis is performed in a vacuum, and the existing literature on the subject matter (no matter how flawed) should influence the study design and the analysis performed. An excellent history of hip arthroplasty can be found in a rather complete reference by Amstutz.[2] The important issue to consider from a historical perspective is the number of different variations in hip arthroplasty design as well as surgical techniques that have been used over time. This evolution includes design changes in both the femoral and acetabular components, material changes to both components, the introduction of modularity, differences in fixation methodologies,[3] and changes in the surgical environment.[2] It is fair to say that there has been a continuum in the evolution of hip arthroplasty, and the changes are still taking place today.

Although much of the existing literature on hip arthroplasty is flawed by the anecdotal nature of the data, the lack of objective outcomes, and the lack of controlled studies, two of points are clear. First, the durability of hip implants in older patients is far superior to that of younger patients.[1, 4-6] In fact, most older patients can be expected to outlive their prostheses.[1, 7] This association is partly explained by the likelihood of greater activity levels in the younger patients.[8]

Second, the results of revision hip arthroplasty are not as promising as with primary arthroplasty, although the degree of difference is questionable. With this in mind it is clear that patient age as well as the type of arthroplasty (primary vs revision) must be accounted for in any analysis.

2.2. What Is the Best Study Design to Use?

It is well known that the best way to compare the efficacy of different medical treatments is through the use of randomized clinical trials. However, although such studies have proven invaluable in the comparison of drug treatments, they are rare in the surgical setting.[9, 10] Aside from the cost factor, the special problems associated with randomized studies in the orthopedic surgery setting have been discussed in the literature.[10, 11] Such issues as surgeon experience with a given prosthesis, postoperative protocol, surgical approach used, the constantly changing surgical environment (such as the earlier hospital discharges of today) make it difficult to know when two patients receiving the same type of prosthesis have actually received the same medical treatment.

There are no accepted long-term randomized clinical trials in the existing hip arthroplasty literature. In the past, there was less of an emphasis on controlled studies, such as randomized clinical trials. The greatest impediment to randomized trials in the hip arthroplasty setting today is the large variety of hip joint replacements presently in use. There are at present well over 100 different hip replacement systems that are in use world-wide. Given the length of time required for a proper comparison of any two components (at least 10 yr), the constant evolution of the component design and surgical technique, and the numbers of such components presently in use, it is now clearly impossible to compare them all in the randomized clinical trial setting.

Thus, we can expect that most of the future publications presenting results of total hip arthroplasty will continue to involve observational studies, where the choice of implant component has not been made by any randomized procedure but by other factors, such as surgeon preference and hospital policy. Although observational studies can always be subject to criticism (as is true of many randomized studies as well), they can also be extremely useful and informative if the analysis is done properly. Observational studies do require extra care in the data analysis and data presentation so that a reasonable comparison of the different types of implants can take place in the literature. All published medical papers should be viewed as comparative in the sense that results of different studies can be compared either informally or formally (in some type of meta analysis).

2.3. What Is the Principal Outcome of Interest?

One of the first issues that must be addressed in any medical intervention study is what criteria will be used to evaluate the effectiveness of the device. All medical interventions are based on certain goals for improving patient condition, even though the expectations of the patient and the surgeon may be different. In the orthopedic literature, the expectations of the surgeon have usually been the basis for evaluating a hip arthroplasty procedure, although the subjective evaluation based on patient outcome questionnaires has recently been introduced.[11]

Regardless of which measurement is used, a failure can always be identified as a case in which the outcome measurement no longer meets a minimum level of acceptance. It is common for treating physicians to categorize patients as "good," "fair," or "poor," based on the outcome measurements used.

A common problem in the orthopedic medical literature is that different authors often use different definitions for failure. Almost all hip arthroplasty surgeons would agree that a hip implant failed if the patient needed revision surgery for such reasons as implant loosening or fracture of the component. However, because of the small number of revision surgeries observed during the first 5 yr following surgery, most of the existing literature concentrates on the surrogate variable of "radiographic implant loosening."[12-16] The definition of implant loosening is based on the extent and magnitude of radiolucent lines in the area of the prosthesis. Unfortunately, the definition for when implant loosening has occurred also varies from author to author.

One problem associated with radiographic definitions of failure is the subjective nature of reading patients' radiographs. A second problem is that patients who are "radiographically loose" frequently show no other clinical signs of failure (such as pain),[14, 15] and the relationship between the surrogate variable (radiographic loosening) and the variable of interest (revision surgery) may not be as immediate as is assumed, and may even depend on other factors. The Kaplan-Meier survivorship curves[17] for THARIES acetabular components, shown in Fig. 1, are based on the time from becoming "radiographically loose" (100% radiolucency around the acetabular component with a minimum width of 1 mm) to the need for a revision of the THARIES

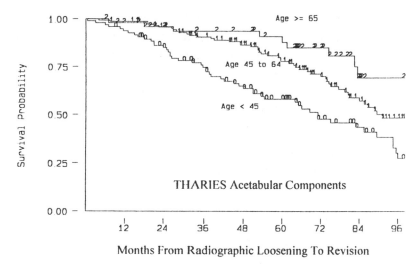

Fig. 1. Kaplan-Meier survivorship curves for time from acetabular loosening to revision surgery based on patient age at surgery. The prosthesis involved is the acetabular component of the THARIES resurfacing prosthesis.

component. These survivorship curves clearly demonstrate that the age of the patient is related to the time of future revision. In particular, few older patients will require a revision despite their being radiographically loose, although 50% of the younger patients will require a revision within 3–4 yr after the appearance of substantial radiolucencies. As mentioned previously, the reduced risk of revision surgery in older patients probably results from their lower activity levels as well as the likelihood of patient death prior to the need for a revision.

2.4. What Should Be the Primary Method of Analysis?

The most unique aspect of the long-term evaluation of any medical intervention is the existence of censored observations; i.e., the time of prosthesis failure is not yet known for all patients. The results of a hip arthroplasty patient who has been followed for 15 yr clearly should not be compared with those of a patient who has received a different prosthesis but has only been followed for 5 yr. In order to account for the differences in the length of time that patients have been followed, the technique of survivorship analysis has been applied to hip arthroplasty.[18–23]

The long-term results of a procedure over time can be presented through the use of a survivorship curve where the percentage of patients continuing to function well is graphed over time. When utilizing survivorship analytic techniques, patients only contribute information about the results during the time period that they have been followed. When the outcome variable used is continuous (such as when quality of life is the major issue) rather than dichotomous (failed vs not failed), the analysis becomes more complicated. One cautionary note concerning all survivorship analysis is that the censoring mechanism is usually assumed to be independent of the risk of failure.

2.5. What Should Be Done if the Present Status of Some Patients Is Not Known?

When providing follow-up for an extended period of time, patients will usually fall into several different categories. Some patients will have been followed until they failed. Other patients may have been followed but have not yet failed at the time of analysis. In an older population, many patients may have died with a well-functioning prosthesis *in situ*. Finally, some patients may be "lost to follow-up" because the present status of their implant is not known because attempts to contact them have been unsuccessful.

Depending on the medical situation, this last group might be very small or very large. In the field of oncology or in situations involving cardiovascular disease (when death is of prime concern), we would expect most patients to keep in close contact with their treating physician, and consequently this situation would not be as much of a problem. However, in elective surgery situations, such as hip arthroplasty, where the initial results are so successful, patients might be less likely to keep in contact with the surgeon who performed the hip arthroplasty. In this situation, a much higher percentage of the patients might become "lost to follow-up" over a 10–15-yr time period. This may also be the case in studies involving medical devices that had already been approved, where effectiveness rather then efficacy was the issue.

The group of patients "lost to follow-up" causes great concern because of the possibility that the reason for the patient being lost may in fact be related to the risk of failure, and consequently, using the usual survivorship analysis techniques may bias the evaluation of the prosthesis. This bias may lead to either an under- or over-estimation of the effectiveness of the implant. In one study, it was shown that patients who were "lost to follow-up" had the same hazard for failing as patients who were not lost.[23] In all studies where there are more than a few patients "lost to follow-up," some additional evidence should be presented to indicate that the independence assumption is reasonable.

2.6. What Should Be Done About Patients Who Died Prior to Requiring a Revision?

In total hip arthroplasty analysis, patients who died with a secure prosthesis present no unusual concern (unlike the situation when evaluating patients with cardiovascular implants, for example) and in most survivorship analysis they should be treated as censored observations. However, the group of patients who died prior to a revision are actually successes of the orthopedic procedure since the goal of the surgery was to have the hip implant survive the lifetime of the patient, not to extend the lifetime of the patient. This situation is referred to as competing risks (implant failure vs patient death) in the statistical literature and it leads to alternative survivorship curves being evaluated rather than the usual Kaplan-Meier curves.[22, 24]

3. Results of the T28/TR28/Charnley Cemented Stem Hip Arthroplasty Component

3.1. Study Design and Sample Characteristics

Given the above considerations of the issues involved in the evaluation of total hip arthroplasty effectiveness, the application of these concepts to the analysis of one observational study is provided below as an example. This study presents results of all total hip arthroplasties (597 hips in 513 patients) performed from 1974 to 1982, in patients who were followed for a period of at least 1 yr and who received a cemented t28 (81%),[5] tr28 (16%), or Charnley prosthesis (3%)[25, 26] as a primary total hip replacement. The year 1974 was chosen as the starting point for the study since after that time no major changes occurred in the operating room environment, the cementing techniques were fairly stabilized, and the use of antibiotics in the cement was routine. All surgeries were performed at the same teaching hospital in the city of Los Angeles. Eighty percent of the surgeries were performed by three surgeons, with seven surgeons performing most of the remaining surgeries.

This is an observational study in that from 1976 to 1982 there was another alternative prosthesis available (the THARIES resurfacing prosthesis) that was used initially in younger, more active patients.

The study is partially prospective in that data were routinely collected on patients and entered into a computer, and also partially retrospective in that a rigorous protocol of patient follow-up was not consistently observed (so the time of patient follow-up was haphazard in many cases) and patient radiographs were evaluated retrospectively.

3.2. Statistical Analysis

The primary method of assessing the effectiveness of the arthroplasty was survivorship analysis. In particular, survivorship curves were drawn using the method of Kaplan and Meier, comparisons between groups were made using log rank statistics, and multivariate analysis was performed using the Cox proportional hazards model.[18, 19, 21-23] In addition, the hazard curves for revision were estimated by fitting a Weibull distribution to the data.[21, 23]

The primary outcome used in this study was failure, as defined by the need for revision surgery. The use of this outcome is objective in that there is no question as to the time when failure occurred (date of the revision surgery). Additional information was also obtained concerning the radiographic evaluation of the long-term implants as well as quality of life information on a subset of patients. Finally, because of the large number of patients falling into the "lost to follow-up" category, an extensive analysis of this subgroup was conducted.

3.3. Evaluation of Possible Bias

In all observational studies, there is a need for extra care to protect against the likelihood that because of patient selection or other factors there might be some bias in the stated results. The first requirement for any presentation in the medical literature is a clear description of the patient population characteristics, including special characteristics of the institutions involved. Only then can the generalizability of the results to a larger population, the validity of comparisons within the paper, or the appropriateness of comparisons with the existing literature be considered.

3.3.1. Patient Population

The patient population in the present study was similar to that of most studies of hip arthroplasty patients. The mean age was 61 ($\pm$ 14) yr at surgery; 60% were over 60 yr and 19% were under 45 yr at surgery. Sixty-nine percent were female. The most common etiologies were osteoarthritis (42%), rheumatoid arthritis (17%), avascular necrosis (11%), and development dysplasia of the hip (7%). In general,

Table 1
Patient Demographics Based on Follow-Up Status

	Patients lost to follow-up	Lost patients subsequently found	All patients in the study
N	234	132	597
Age, yr	60 ($\pm$ 12)	61 ($\pm$ 12)	61 ($+$ 14)
Males, %	33	40	31
Weight, kg	79 ($\pm$ 19)	82 ($\pm$ 16)	82 ($\pm$ 17)
Female, %	67	60	69
Weight, kg	68 ($\pm$ 18)	69 ($\pm$ 16)	67 ($\pm$ 17)
Etiology			
Osteoarthritis, %	39	51	42
Avascular necrosis, %	12	6	12
Dysplasia of the hip, %	6	7	7
Rheumatoid arthritis, %	17	17	17

the population was not exceptionally active, because many patients who desired to participate in sporting activities received the THARIES prosthesis during that time period.

3.3.2. Evaluation of Patients "Lost to Follow-Up"

In the present study, 234 patients had not been seen during the previous 3 yr or longer at the beginning of the analysis and could not be contacted by telephone. With such a large percentage of patients in this "lost to follow-up" category (46%) the authors must justify the usual independence assumption required for traditional survivorship analysis. An earlier study from this institution suggested that the assumption of independence may have been reasonable, but because of the large number of patients "lost to follow-up," some additional evidence was clearly indicated.

Any study with more than just a few patients "lost to follow-up" should compare the patient demographics of those patients with the original group of patients. As shown in Table 1, there is no suggestion that this group of 234 patients was systematically different from the entire group (columns 1 and 3). However, in this instance an additional (and time-consuming) telephone search was initiated based on information in the patient charts that allowed 132 of these 234 patients to be found. The results for this new group of patients were then compared to the results that would have been obtained at the start of the study without any additional information.

The 5- and 10-yr prosthesis survivorship of the original vs recently found groups of patients were very similar (99% vs 100% at 5 yr, and 94% vs 97% at 10 yr, whereas at 15 yr the 132 patients who were recently found had better results; 90% vs 77%). Based on these results, the information on these 132 patients was updated in the original data set, and the survivorship analysis proceeded with an understanding that the actual long-term results (if all patients were to be contacted) might be slightly better than those that were presented.

3.3.3. Possibility of Patient Selection Bias

Beginning in 1976 a new resurfacing prosthesis called the THARIES, with no stem in the femoral component, was introduced and subsequently used routinely for younger, more active patients. Thus, there is a strong possibility that results from the patients in this study would not be comparable to the results that would have been obtained if all patients had received the cemented stemmed prosthesis. In order to evaluate this possibility, the patients were divided into three groups corresponding to three time periods:

1. 1974–1975, prior to the introduction of the THARIES;
2. 1976–1979, a period of substantial THARIES use in younger patients; and
3. 1981–1982, declining use of the THARIES.

The estimated prosthesis survivorship for these time periods was 97, 95, and 96% for Groups 1, 2, and 3, respectively, at 10 yr; and 86 and 81% for Groups 1 and 2, respectively, at 15 yr (Group 3 has not yet been followed for 15 yr). Thus, there is some evidence that any possible selection bias is small in this case, and presuming that these results are representative of the general population seems reasonable.

Although the analyses presented here and previously do not prove without a doubt that there might not be some bias, they do give the reader some indication that any such bias, if it exists at all, is small and not clinically significant. Of course, some readers may discount the results of all observational studies because of the potential for bias. However, even in randomized studies, the possibility of existing bias between the comparison groups must be investigated, since randomization only protects against bias "on average." Furthermore, randomized studies also suffer from the possibility of patient selection bias because of the requirement of patient consent prior to entry into the study.

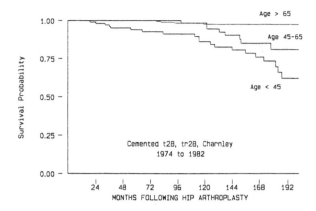

Fig. 2. Kaplan-Meier survivorship curves for time from arthroplasty surgery to aseptic revision surgery based on age at surgery. The prostheses involved were the cemented stemmed t28, tr28, and Charnley.

3.4. Analysis of Prosthesis Survival

The existing literature suggests that unless the population has a homogeneous age distribution (all patients very young or old at the time of surgery), it is meaningless to present combined results of total hip arthroplasty for the entire population. It is well documented that the age of the patient at surgery is an important indicator of the risk for revision surgery and consequently any presentation of hip arthroplasty results must take into account the age of the patients. Thus, the first approach in the analysis is to examine the effect of age on the risk for revision surgery as well as to look for other important prognostic factors that are predictive of early revision surgery.

The Kaplan-Meier survivorship curves for time to aseptic revision for any reason are shown in Fig. 2. The effect of age at surgery is obvious with a 15-yr prosthesis survivorship of 96% in patients over 60 yr at surgery and only 63% in those patients under 50 yr at surgery. The survivorship curves based on age are both clinically and statistically significant ($p < 0.0001$, log rank test).

Other variables investigated include gender, weight, etiology, year of surgery, and the presence of bilateral hip implants. The statistical tests used were the log-rank test for categorical independent variables (such as gender and etiology) and the Cox proportional hazards model

Table 2
Interaction Among Patient Age,
Etiology of Osteoarthritis, and Revision Surgery
15-Yr Prosthesis Survivorship Estimates with Standard Errors

Population	Osteoarthritis	Nonosteoarthritis
Survivorship estimates, %		
Entire population	95 ($\pm$ 2.2)	83 ($\pm$ 3.6)
Age < 50 at surgery	100	72 ($\pm$ 8.0)
Age 50–60 at surgery	86 ($\pm$ 6.2)	79 ($\pm$ 8.0)
Age > 60 at surgery	97 ($\pm$ 1.7)	99 ($\pm$ 1.2)

for continuous variables, such as patient weight. Of these variables, only an etiology of osteoarthritis ($p < 0.005$, log rank test) was found to be related to time of aseptic revision of the prosthesis. However, since most osteoarthritis patients are older, the actual relationship between this etiology and revision is not clear.

3.4.1. Interaction of Prognostic Variables

Whenever several prognostic variables are found to be related to the outcome variable they should always be examined in a multivariate way, especially considering the possibility that there may be some interactions. For example, prior to concluding that osteoarthritis places the patients at reduced risk over other etiologies, we should determine if the association between osteoarthritis and time to revision noted above is true for patients in all age categories. It is clear from Fig. 2 that there are more likely to be differences between subgroups of young patients rather then older patients since the results for all older patients are favorable. If the sample size is large enough, the easiest way to compare the effect of two or more variables is through stratification.

Presented in Table 2 are the 15-yr results comparing the osteoarthritis and nonosteoarthritis patients in the entire group as well as in the three age categories: under 50 yr, 50–60 yr, and over 60 yr at surgery. These results suggest that the relationship between the etiology of osteoarthritis and revision surgery results from a difference in the younger patients rather than in the older patients. Evaluation of prognostic factors is incomplete unless interactions with other important variables are also considered.

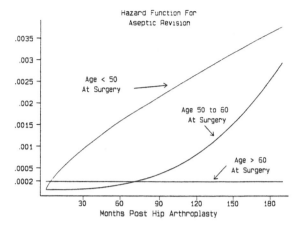

Fig. 3. Hazard curves for time from surgery to aseptic revision based on the Weibull distribution. Because of the lack of sufficient number of revisions, the hazard for the patients over 60 yr at surgery is based on the exponential distribution.

3.4.2. Investigation of the Hazard Rates

When performing survivorship analysis, the "hazard rates" (instantaneous risk of failure) over time should be investigated. In particular, it is of interest to know if the risk of failing increases or decreases over time. One approach to this is to fit a parametric survivorship function (such as the Weibull distribution) to the data and evaluate the hazard based on that model.[21, 23] The estimated hazard rates based on the age of the patients at surgery are presented in Fig. 3. It is clear that the younger patients have a steadily increasing (almost linear) hazard whereas the middle age group (50–60 yr) has a minor initial hazard that rapidly increases over time. The older group has a suggested very minor and constant hazard. These hazard functions can be explained by the possibility of component wear related to the different patient activity levels that are expected based on the age of the patients.

3.4.3. Investigation of Complications

One of the most important issues in elective surgical procedures, such as hip arthroplasty, is the serious complication rate. Serious complications from hip arthroplasty surgery are infrequent today. This is partially because of changes in the surgical environment, such as the use of laminar flow in the operating room[3] and antibiotics in the cement, reducing the complication rate of sepsis. In the present

Table 3
Long Term Radiographic Results by Age

	Patient age at surgery, yr		
	< 50	50–60	> 60
N	35	22	32
Mean follow-up time, mo	127	140	127
Wear, mm	1.7 (± 1.4)	1.5 (± 0.9)	0.9 (± 0.8)
Radiographically loose, 100% lucency with max > 1 mm	54%	50%	28%

study, 1.6% of the surgeries were associated with in-hospital complications. Nerve palsies occurred in 1.1%, hip dislocations in 0.4%, and pulmonary emboli in 0.2%. All patients in this series participated in a Warfarin protocol designed to reduce the rate of pulmonary emboli. There were no deaths associated with pulmonary emboli in this series, although one death has subsequently occurred to a patient who was also on a Warfarin protocol.

There were three patient deaths within 6 mo of surgery, and the possibility that these deaths were partly related to the surgical procedure can not be eliminated. The occurrence of sepsis required a revision of 0.2% of prostheses over the first 10 yr. The risk of revision because of sepsis was 2.5 times as likely in patients 60 yr or younger than in patients over the age of 60 at surgery ($p < 0.082$).

3.4.4. Radiographic Evaluation of Long-Term Survivors

Wear of the acetabular liners is now considered to be directly related to risk of aseptic revision.[27] In this study, there was a clear association between wear and age at surgery. In addition, the presence of 100% radiolucencies of width 1 mm or more as observed in a higher percentage of younger versus older patients (*see* Table 3). This corresponds with the observed increased risk of revision in the younger population. There is little doubt that a difference in patient activity following hip arthroplasty is largely responsible for the poorer results in the younger patients. An earlier study involving a different type of component (the THARIES resurfacing) also found an association between patient activity and the need for patient revision.[8]

3.4.5. Patient Death

The death of the patient takes on a different meaning in the orthopedic vs the oncology or cardiovascular setting. The death of the

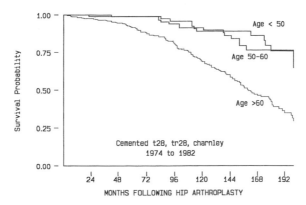

Fig. 4. Kaplan-Meier survivorship curves for time from arthroplasty surgery to patient death based on age of patients at surgery.

patient prior to revision of the implant represents a situation where failure of the orthopedic device is impossible. This leads to the possibility of estimating a different survivorship curve then the usual Kaplan-Meier curve.

The Kaplan-Meier curve is a conditional survivorship curve because it estimates the probability that the prosthesis will not fail among those patients who will actually survive until the given time period. In situations in which there would be few patient deaths, this conditional aspect is not important. However, in situations, such as an older patient population, where many patients might die prior to requiring a revision surgery, a better survivorship probability to estimate is the probability that the prosthesis will survive the patients' lifetime.[22, 24, 25]

Figure 4 displays the survivorship of the patient population based on age category at surgery. The greater risk of patient death in the older patient partially explains why the prosthesis survivorship is far superior in that group. If a given prosthesis is designed to last only 20 yr, but the patient dies prior to 20 yr, then there will be no prosthesis failure. In this situation, patient death is considered a competing risk to prosthesis failure; a prosthesis designed to last 30 yr is not needed for patients who will only live for 20 yr or less.

The methodology for estimating the probability of prosthesis survival over the patient's lifetime is given in Korn[24] and Kalbfleisch.[22] From our data, the 15-yr estimate of aseptic revision occurring during the patient's lifetime is 3.4% (estimated 95% confidence interval 2–7%).[27] This 15-yr estimate with narrow confidence intervals (relative to those of the Kaplan-Meier curve) suggests that in the older patient the cemented stemmed prosthesis is very durable and unlikely to be surpassed with additional modifications in the future. When the purpose of the medical implant is not related to extending the patient's lifetime, then estimating the probability of implant survival during the patient's lifetime should be considered as a possible alternative to the usual Kaplan-Meier survivorship curve.

4. Conclusion

The use of these earlier cemented hip arthroplasty components has been shown to be very durable in older patients. Thus, any decision regarding which prosthesis to use in this population will likely be based on factors other than improved prosthesis durability. All studies involving hip arthroplasty should present results using the technique of survivorship analysis either stratified by patient age or with age as a covariate in a multivariate statistical technique, such as the Cox proportional hazards model. Observational studies rather then randomized studies will continue to be used in evaluating the effectiveness of total hip arthroplasty, and the proper and careful statistical analysis of the results will be more important than ever.

References

1. Kay, B., Dorey, F., Johnston-Jones, K., Cracchiolo, A., Amstutz, H., and Finerman, G. 1995. Long-term durability of cemented primary total hip arthroplasty. *J. Arthroplasty* 10:S29–S38.
2. Amstutz, H. 1991. *Hip Arthroplasty*. Churchill Livingstone, New York.
3. Charnley, J. 1964. A clean-air operating enclosure. *Br. J. Surg.* 51: 202–205.
4. Dorr, L. D., Luckett, M., and Conaty, J. P. 1990. Total hip arthroplasty in patients younger than 45 years: a nine to ten year follow-up study. *Clin. Orthop.* 260:215–219.
5. Amstutz, H. C. 1973. Trapezoidal-28 total hip replacement. *Clin. Orthop.* 95:185–167.
6. Chandler, H. P., Reineck, F. T., Wixson, R. L., and McCarthy, J. C. 1981. Total hip replacement in patients younger than thirty years old. *J. Bone Joint Surg. Am.* 63:1426–1434.
7. Schulte, K. R., Callaghan, J. J., Kelley, S. S., and Johnston, R. C. 1993. The outcome of Charnley total hip arthroplasty with cement after a minimum of twenty-year follow up. The results of one surgeon. *J. Bone Joint Surg. Am.* 75:961–975. [Published erratum appears in *J. Bone Joint Surg. Am.* 1993. 75:1418.]
8. Kilgus, D., Dorey, F., Finerman, G., and Amstutz, H. 1991. Patient

activity, sports participation, and impact loading on the durability of cemented total hip replacements. *Clin. Orthop.* 269:25–31.

9. Bland, M. 1995. *An Introduction to Medical Statistics.* Oxford Medical, New York.
10. Rudicel, S. and Esdaile, J. 1985. The randomized clinical trial in orthopaedics. Obligation or option? *J. Bone Joint Surg. Am.* 67:1284–1293.
11. Keller, R., Rudicel, S., and Liang, M. 1994. Outcomes research in orthopaedics. In *Instructional Course Lectures,* (Schafer M. ed.) Vol. 43. American Academy of Orthopaedic Surgeons.
12. DeLee, J. G. and Charnley, J. 1976. Radiological demarcation of cemented sockets in total hip replacement. *Clin. Orthop.* 121:20–32.
13. Gruen, T. A., McNeice, G. M., and Amstutz, H. 1979. Modes of failure of cemented stem type femoral components: a radiographic analysis of loosening. *Clin. Orthop.* 141:17–27.
14. Hodgkinson, J. P., Shelley, P., and Wroblewski, B. M. 1988. The correlation between the roentgenographic appearance and operative findings at the bone-cement junction of the socket in Charnley low friction arthroplasties. *Clin. Orthop.* 228:105–109.
15. Carlsson, A. S. and Gentz, C. F. 1984. Radiographic versus clinical loosening of the acetabular component in noninfected total hip arthroplasty. *Clin. Orthop.* 185:145–150.
16. Schmalzried, T.P., Kwong, L.M., and Jasty, M. 1992. The mechanism of loosening of cemented acetabular components in total hip arthroplasty. *Clin. Orthop.* 274:60–78.
17. Kaplan, E.I. and Meier, P. 1958. Nonparametric estimation from incomplete observations. *J. Am. Stat. Assoc.* 53:457–481.
18. Lee, E. 1980. *Statistical Methods For Survival Data Analysis.* Lifetime Learning, Belmont, CA.
19. Dobbs, H. S. 1980. Survivorship of total hip replacements. *J. Bone Joint Surg. Br.* 62:168–173.
20. Cox, D. R. and Oakes, D. 1984. *Analysis of Survival Data.* Chapman and Hall, New York.
21. Dorey, F. and Amstutz, H. 1986. Survivorship analysis in the evaluation of joint replacement. *J. Arthroplasty* 1:63–69.
22. Kalbfleisch, J. D. and Prentice, R. L. 1980. *The Statistical Analysis of Failure Time Data.* John Wiley, New York.
23. Dorey, F. and Amstutz, H. 1989. The validity of survivorship analysis in total joint arthroplasty. *J. Bone Joint Surg. Am.* 71:544–548.
24. Korn, E. and Dorey, F. 1992. Applications of crude incidence curves. *Stat. Med.* 11:813–829.
25. Charnley, J. 1961. Arthroplasty of the hip. A new operation. *Lancet* 1:1129–1132.
26. Wroblewski, B. M. 1988. Wear and loosening of the socket in the Charnley low-friction arthroplasty. *Orthop. Clin. North Am.* 19:627–630.
27. Dorey, F. and Korn, E. 1987. Effective sample sizes for confidence intervals for survival probabilities. *Stat. Med.* 6:679–687.

8

Injectable Collagen and a Rare Adverse Event— True Association or Artifact?
Results of Postmarket Surveillance Research

Diane E. Mandell, Rosanne B. McTyre, Frank DeLustro, and Ross Erickson

1. Introduction

Collagen-based medical devices have been widely used in human soft tissue ranging from suture material to injectable dermatologic implants, and have a long history of safe and effective use.[1] Zyderm® and Zyplast® implants are forms of injectable bovine collagen (highly purified bovine dermal collagen, containing at least 95% Type I and < 5% Type III collagen in a fibrillar suspension of physiologic buffered saline and 0.3% lidocaine). Injectable collagen (Zyderm/Zyplast) is indicated for use in correction of contour deformities of the dermis in nonweight-bearing areas such as correction of distensible acne scars; atrophy from aging, disease, or trauma; and other soft tissue defects. Local complications, as determined in prospective trials of the safety and effectiveness of this device, include the low risk of infection that accompanies any transcutaneous procedure and a low level of a local hypersensitivity response (1–2% of treated patients). Anecdotal reports received postmarketing suggested that injection of Zyderm/Zyplast implants may be associated with an increased risk of a rare autoimmune disease, polymyositis/dermatomyositis (PM/DM).

From: *Clinical Evaluation of Medical Devices: Principles and Case Studies*
Edited by K. B. Witkin Humana Press Inc., Totowa, NJ

In addition, a report by Cuckier et al. (1993)[2] suggested that there may be a positive association between bovine collagen dermal implants and PM/DM. However, a critical evaluation of the available data by experts assembled by the Food and Drug Administration (FDA) concluded that there was "...insufficient statistical or biological evidence to support a conclusion that collagen injections cause autoimmune or connective tissue diseases in persons without a history of these diseases...."[3]

Despite both the positive outcome of the FDA experts meeting and the large bodies of preclinical data and clinical experience demonstrating the safety and effectiveness of this device, Collagen Corporation chose to directly address the new safety issues raised for the bovine collagen-containing materials Zyderm and Zyplast. First, a careful evaluation of available epidemiological studies on the potential association between collagen injection and PM/DM[2, 4, 5] was undertaken to identify design flaws that rendered the conclusions suspect. One of the greatest weaknesses of these epidemiological evaluations was the uncertainty in the estimated duration of patient follow-up after bovine collagen exposure. Duration of patient follow-up is an integral factor in the calculation of the number of expected cases of PM/DM among the collagen-treated population. Since duration of follow-up is readily audited from available patient records, the sponsor initiated a doctor audit survey to more accurately determine the patient follow-up duration. The experimental approach and the results of the postmarket survey are described in this chapter.

2. Background

Two epidemiological studies in the published literature directly address the question of a potential link between bovine collagen implants and the autoimmune diseases PM and DM.[2, 4] In the historical cohort study by Cukier et al., nine subjects who were injected with bovine collagen implants were reported as having developed inflammatory myositis (PM or DM) an average of 6.4 mo after treatment, from among an estimated 345,000 total patients treated between the years 1980 and 1988.[2] The authors examined the association between bovine collagen injection and the development of PM/DM by comparing the reported cases of PM/DM to the age-, race-, and gender-specific incidence of myositis in the general population (obtained from a hospital-based study in Allegheny County, PA, conducted by Oddis et al.[6]). Cukier et al. calculated a statistically significant increase in the observed incidence of PM/DM in collagen-treated subjects compared to the number of cases expected based on the incidence of PM/DM in the general population. On the basis of these results, the authors concluded that the risks vs the benefits of using collagen implants are greater than previously believed and should be reassessed.

Several shortcomings in the Cukier et al. analysis are noteworthy. Most importantly, the definition of what constituted a case differed between the observed and expected groups. Cukier et al. identified nine alleged PM/DM cases who were either treated with collagen ($n = 7$) or who had been skin-tested only ($n = 2$). Although these cases fulfilled some of the criteria for classsification of PM/DM, they were not confirmed cases (i.e., clinical testing for PM/DM was performed only in some subjects). As Hochberg[7] indicated in an editorial on the Cukier et al. study, the expected cases were based on a population at risk obtained from company sales reports that included only individuals who had received collagen treatments; those who received skin tests were not included. In addition, there were other potential sources of underestimation of expected numbers in this study, including assumptions regarding the background incidence rate of PM/DM used in the calculation (specifically, the generalizability of this background rate to a nonhospital population). In conclusion, noncomparability between the numerator (which was overestimated) and the denominator (which was underestimated) lead to an overall erroneous overestimation of the risk associated with collagen implants in this study, that is, to a potential association where one did not exist.

In an analysis of the potential relationship between use of collagen implants and PM/DM, Rosenberg and Reichlin also compared the number of observed cases of PM/DM in subjects who had been treated with collagen to the number of PM/DM cases expected.[4] Seven patients met the criteria for classification of PM/DM using all available sources of information (i.e., manufacturing surveillance system, the media, registry data, and letters to US rheumatologists). The expected number of cases was estimated based on the population at risk (600,000 subjects injected with collagen, as provided from manufacturing data), the duration of follow-up (as estimated by 109 physicians who used collagen, then extrapolated using modeling techniques), and the most recent background incidence of Oddis et al., adjusted for the last 5 yr of the 20-yr study period (1978–1982). Based on an incidence rate of 10.2 patients/million/yr, 13 expected cases were calculated compared to the 7 cases observed, a statistically significant difference. Sensitivity analysis suggested that, using the most conservative estimates, 12 cases would be expected and by using worst-case assumptions, 130 cases would be expected. These data did not support an association between collagen implants and PM/DM. The previous estimates of observed and expected rates of PM/DM have been largely based on assumptions regarding the collagen-treated population (the population at risk for developing PM/DM) and the duration of follow-up, and may have included individuals who were not confirmed as having definite PM/DM in the observed cases.

3. Collagen Corporation's Postmarket Doctor Audit Survey

The objective of the 2-yr postmarket doctor audit survey conducted by Collagen Corporation was to measure the actual duration of follow-up for collagen-injected subjects and calculate the expected rate of occurrence of PM/DM in subjects treated with injectable collagen in the form of Zyderm or Zyplast. Incorrect estimates for follow-up duration could be responsible for a large margin of error in the calculation of the expected occurrence, which would make a comparison to the observed occurrence of PM/DM in collagen-treated subjects difficult or inaccurate.* This postmarket doctor audit survey improves on previous studies by more clearly characterizing the observed cases and computing the actual duration of follow-up. Accurate determination of duration of follow-up is a critical element for estimating the expected cases of PM/DM among the collagen-treated population.

In order to assess the observed vs the expected occurrence of PM/DM, the following "critical epidemiological factors" must be accurately estimated:[8]

---

* The number of expected cases of PM/DM among the collagen-treated subjects is calculated by multiplying the estimated person-years-at-risk for each age, gender, and race subgroup with the respective background incidence rate. Person-years-at-risk is computed by multiplying the number of new cases of PM/DM per year and the estimated duration of patient follow-up.

1. The background rate of occurrence of PM/DM must be assessed in a reference population (i.e., without collagen).
2. The number of cases of PM/DM in collagen-treated patients must be determined.
3. The total number of subjects treated with injectable bovine collagen implants must be accurately evaluated.
4. The duration of patient follow-up after injection must be determined accurately, with an adequate latency for PM/DM to develop.

Accurately determining the duration of follow-up was the primary objective of this patient chart audit, using accurate estimates for the background rate of PM/DM and the total number of subjects treated with collagen from previous analyses.

3.1. Methods

The study methodology was recently published in the scientific literature by Hanke et al.[8] and is presented again below.

3.1.1. Recruitment of Physicians

Physician recruitment was as follows: 4469 physicians in the United States and Canada who purchased injectable collagen between July 1987 and June 1988 were stratified according to the size of their practice in order to represent subjects from large, medium, and small practices equally. From this group, 1474 physicians were randomly selected and recruited by mail according to stratification weights: 100% of large practices, 50% of medium practices, and 25% of small practices were recruited, with the goal of enrolling the 1000 subjects necessary to achieve at least 90% statistical power for a two-tailed binomial distribution (the probability of developing the disease). Physicians recruited from the first letter mailed comprised the first study group. To minimize the impact of nonresponse from physicians (it was hypothesized that physicians with longer follow-up times may respond more favorably to the request), a second group of physicians was recruited from those that did not respond to the first letter.

3.1.2. Patient Records Audited

A two-step sampling scheme was employed to select patient records, with a sampling of patient records proportionate to practice size. Specifically, patient records were carefully sampled to minimize the impact of practice size and collagen use; subjects were inversely selected, 25% of subjects from large practices, 50% from medium practices, and 100% from small practices.

All record audits were conducted on-site at each physician's office. Auditors compiled a complete list of all patients treated with collagen implants for the audit period through review of the physician's billing records, appointment books, and/or other sources. All subjects had soft tissue augmentation with injectable bovine collagen implants produced by Collagen Corporation, excluding those subjects who had been skin-tested but not treated with collagen. Follow-up interval data were abstracted from 100% of charts from collagen-treated patients. These data included demographic information, the date of the first and last bovine collagen injection, and the date of the last office visit for any reason.

3.1.3. Duration of Follow-Up

The duration of follow-up was then calculated from the patient chart audit as the time interval between the first collagen implant and the last office visit for any reason. The rationale for this is that soft tissue augmentation therapy using injectable collagen is typically an ongoing replacement therapy rather than a single treatment.

3.1.4. Observed Cases of PM/DM

Fifteen alleged cases of PM/DM had been reported to the manufacturer; the actual number of observed cases was determined in a previous study using the clinical diagnostic criteria of Bohan and Peter;[9] these were considered the observed cases for this analysis.

3.1.5. Expected Cases of PM/DM

The expected number of cases of PM/DM is calculated as: person-years-at-risk (for each age, gender, and race subgroup) × the appropriate background incidence rate.

As stated earlier, the person-years-at-risk is calculated as: the collagen-treated population (the population at risk) per year × the follow-up duration. For this analysis, the background incidence rate (for specific age, gender, and race subgroups) was used as determined by Oddis et al.,[6] who determined that over a 20-yr observation period, there was an increasing rate of PM/DM diagnosis or occurrence over time. The 5-, 10-, and 20-yr background incidence rates were used to calculate the expected numbers of cases of PM/DM. The collagen-treated population per year was determined by manufacturing sur-

Table 1
Distribution of Subjects by Geographic Region
and Physicians' Collagen Practice Size[a]

Collagen practice size	Number (%) of 2622 subjects, by region				
	Northeast	Midwest	South	West	Total by size
Small	189 (7.2)	114 (4.4)	174 (6.6)	299 (11.4)	776 (29.6)
Medium	152 (5.8)	139 (5.3)	198 (7.6)	289 (11.0)	778 (29.7)
Large	184 (7.0)	76 (2.9)	475 (18.1)	333 (12.7)	1068 (40.7)
Total	525 (20.0)	329 (12.6)	847 (32.3)	921 (35.1)	2622 (100.0)

[a] Reprinted with permission from ref. 8.

veillance data, reports arising from media coverage, registry data, and letters to US rheumatologists. The duration of follow-up was calculated using the patient chart audit in this study.

3.2. Results

3.2.1. Recruitment of Physicians

Sixty-seven physicians responded to the first letter of recruitment and comprised the first study group; records from 1966 subjects were audited from this group. A second study group of 12 physicians (with records from 656 subjects audited) was recruited from the initial group of nonresponders in order to reduce physician selection bias. For most analyses, data from subjects were pooled from both study groups (2622 subject records total).

3.2.2. Subject Records Audited

Geographic region and practice-size distribution of the 2622 subjects are shown in Table 1. Thirty-one US states and three Canadian provinces are represented. Ninety-eight percent of subjects were Caucasian, 93% female. The age at treatment ranged from 15 to 85 yr, with the mean (± SD) age equal to 46 ± 12 yr.

3.2.3. Duration of Follow-Up

The overall mean duration of follow-up as determined from patient chart audit was 4.0 ± 2.8 yr. Mean values for the first and second physician study groups were 3.8 ± 2.8 (1966 subjects) and 4.6 ± 2.8 yr (656 subjects), respectively. Table 2 shows the average years of follow-up for the 2622 subjects according to practice size and geographical region and confirms the overall reported duration of follow-up of 4 yr.

Table 2
Average Number of Years of Follow-Up (± Standard Deviation)[a]

Collagen practice size[b]	Region[b]				Overall average by size
	Northeast	Midwest	South	West	
Small	3.3 ± 2.8	4.1 ± 2.9	3.6 ± 2.9	3.7 ± 2.8	3.7 ± 2.8
Medium	3.7 ± 3.0	3.8 ± 2.8	3.7 ± 2.8	4.2 ± 2.9	3.9 ± 2.9
Large	4.1 ± 2.8	5.0 ± 2.9	4.3 ± 2.6	4.6 ± 2.7	4.4 ± 2.7
Overall average	3.7 ± 2.9	4.2 ± 2.9	4.0 ± 2.7	4.2 ± 2.9	4.0 ± 2.8

[a] Reprinted with permission from ref. *8.*
[b] $p < 0.01$; Large practice area size > medium; large > small; West region > Northeast. The total sample consisted of 2622 subjects.

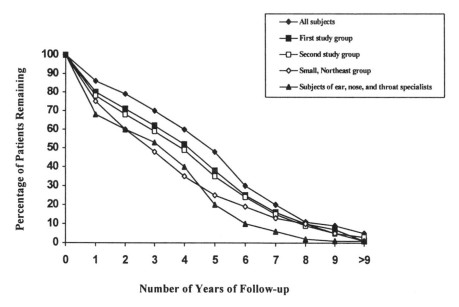

Fig. 1. Subject follow-up patterns for various physician subgroups.

Figure 1 shows the follow-up patterns for the various physician sub-groups, including all subjects ($n = 2622$); the first study group ($n = 1966$); the second study group ($n = 656$); a small, Northeast group ($n = 189$); and subjects of ear, nose, and throat specialists ($n = 105$). This figure demonstrates a similarity in patterns of follow-up regardless of the study group, geographics, or underlying condition.

Table 3
Expected Number of Cases of PM/DM with Variation
in Incidence Rate and Duration of Follow-Up, 1981 Through June 1993

PM/DM incidence rates for reference population[a]	Observed average years of follow-up from record audit[b]	Expected number of cases among population-at-risk[c]	Number of cases for statistical significance
5-yr: 10.2	3.3	26.6	36
	4.0	30.2	40
	5.0	35.8	47
10-yr: 8.5	3.3	24.8	34
	4.0	28.7	39
	5.0	33.3	44
20-yr: 5.5	3.3	15.6	23
	4.0	17.7	26
	5.0	21.0	30

[a]Three estimates of the background rate of PM/DM were used from data published by Oddis et al. (1990).[6] The background rates used were the most recent 5-yr rate of the 20-yr study, the most recent 10-yr rate, and the 20-yr rate.

[b]Duration of follow-up, as determined in this study: The shortest observed duration of follow-up (3.3 yr), the overall mean (4.0 yr), and the longest observed duration of follow-up (5.0 yr) are shown (Table 2).

[c]Calculated as the person-years-at-risk (determined using the observed demographics, the probability of follow-up, and the number of subjects treated each year from 1981 to 1993) multiplied by the background incidence rate of PM/DM (Oddis et al.).[6]

3.2.4. Observed and Expected Cases of PM/DM

Table 3 shows the impact on the expected number of cases of using different background rate estimates for PM/DM (the most recent 5-yr rate of the 20-yr study, the most recent 10-yr rate, and the 20-yr incidence rates from the Oddis et al. data). Using these rates together with different estimates for the duration of follow-up as determined in this study by subject chart audit (the overall mean follow-up time, and the shortest and longest observed duration of follow-up from Table 2), results in different estimates for the expected number of cases of PM/DM. The expected number of cases is shown to range from 15.6 to 35.8 depending on the parameters used; a more than twofold difference.

The observed rate of occurrence of PM/DM had previously been determined to be seven individuals out of 15 who were reported to the

Table 4
Observed vs Expected Number of Cases of PM/DM[a]
1981 Through June 1993

Estimated collagen-treated population (by race/gender subgroup)[b]	Observed (number of cases)	Expected (number of cases)	Number needed for statistical significance[c]
White female	6	26.8	37
White male	1	0.90	4
Nonwhite female	0	2.40	6
Total	7	30.2	40

[a] Calculations for the observed and expected number of cases by gender and race are based on the most probable assumptions: Oddis et al. (1990)[6] 5-yr data and 4.0 mean years of follow-up.

[b] Nonwhite male subjects were excluded because they comprised $< 1\%$ of the collagen-treated population.

[c] Poisson distribution ($p < 0.05$).

manufacturer and who fit the criteria for probable or definite PM/DM. The observed cases are shown in Table 4 by gender and race. This table also shows the expected numbers of cases of PM/DM. Based on the most probable assumptions (Oddis et al. 5-yr data and 4.0 mean observed years of follow-up) and an exact one-sided Poisson distribution, $p < 0.05$, the overall expected number of cases of PM/DM for the period 1981–1993 is 30.2, and the number of cases needed for statistical significance is at least 40.

3.3. Discussion

The alleged relationship between injectable collagen implants and PM/DM is not supported by biological evidence. Animal models have been used for collagen-induced arthritis and autoimmune disorders from Types II, XI, and IV collagen; however, current data do not suggest that this association is true for Types I (the principle component of Zyderm/Zyplast) or III collagen.[10–12] Therefore, there is no biological evidence to support an association between Type I or III collagen and immunological dysfunction.

Clinical data regarding observed and expected PM/DM have been scrutinized with respect to using flawed assumptions. The 15 alleged observed cases of PM/DM that were reported to the manufacturer have been evaluated against the clinical diagnostic criteria of Bohan and Peter.[9] Of these 15 cases, only seven were found to fit the criteria for probable or definite PM/DM. Six of these were white females and one was a white male. There is a high degree of certainty that the manufacturer has successfully determined the number of cases of PM/DM in the collagen-treated patient group, because case-finding has reached beyond passive reporting to the manufacturer. PM/DM case reports were solicited from the American Rheumatism Medical Information (ARAMIS) database, readers of the scientific journal *Arthritis and Rheumatism*, and the mailing list of the American Rheumatism Association.[5] Allegations of an association between collagen injection and the onset of PM/DM were publicized on television news and talk shows ("Cable News Network," "20/20," "Sally Jesse Raphael") and in the newspaper. No new cases have been identified despite this widespread publicity.

Earlier statistical estimates of the expected cases of PM/DM were limited by using faulty estimates for three critical epidemiological factors (background rate of PM/DM, numbers of subjects treated, and duration of follow-up). Erroneous assumptions for these factors would result in a miscalculated number of expected cases of PM/DM among treated individuals. In previous calculations, statistical significance for an association between collagen injection and PM/DM has only been achieved when the most conservative estimates for all three parameters have been used. The impact of using different estimates for these critical epidemiological factors is described below.

Previous epidemiological studies have estimated the background rate of occurrence of PM/DM to range from 2.2 to 5.5 new cases/million population.[6, 13, 14] The impact of using different background rates is shown in Table 3, where calculating expected cases varies as much as twofold as a result of using different background rates (e.g., using the 5-yr background rate vs the 20-yr rate with the shortest observed follow-up time results in 26.6 and 15.6 expected cases of PM/DM, respectively).

For the present calculation of expected rates of occurrence of PM/DM, background incidence data from the study by Oddis et al. was chosen because it is the most recent and demographically representative study in which the incidence of PM/DM is determined by age, gender, and race for different periods of time over 20 yr. In the Oddis et al. study, rates of occurrence were based on hospital diagnoses. The estimate from a hospital population is probably greater than the actual rate in the general population and was therefore conservative for the purposes of calculating expected PM/DM. Oddis et al. determined that 5.5 new cases of PM/DM occurred per million population over 20 yr and 8.5 cases per million occurred during the second 10-yr period (1973–1982). The rate for the second 10 yr was approximately three times greater than the rate for the first 10 yr (1963–1972). Because these data suggest that detection of PM/DM has improved over time, incidence rates for the most recent 5-yr increment (1978–1982) of the published 20-yr data were used to determine the expected incidence of PM/DM among the treated population.

The number of subjects treated with bovine collagen was provided by the manufacturer from their records and was published previously.[7] A conservative estimate was used for calculations in this study, although the magnitude of the underestimate is not certain.

The duration of follow-up used in previous calculations has ranged from 1 to 3 yr. Varying the duration of follow-up also has a great impact on the expected cases of PM/DM calculated, as illustrated in Table 3 (e.g., using the 5-yr background rate with the shortest vs the longest follow-up time results in 26.6 vs 35.8 expected cases of PM/DM, respectively).

The present study improves on the follow-up estimates used by others; subject charts were audited to actually determine the duration of subject follow-up. The overall mean follow-up duration was determined to be 4.0 yr, the shortest estimate 3.3 yr, and the longest estimate 5.0 yr. (Table 2). The follow-up duration determined in this study accurately represented subjects treated with collagen because of a number of techniques used to minimize sampling bias: A large sample size increased the statistical power of the measure, physicians were randomly sampled, physicians were stratified by the size of their practice, and a wide geographic area was represented.

Observed and expected rates of PM/DM were calculated using the most probable assumptions, as described in this chapter. Using the Oddis et al. 5-yr data, 4.0 mean years of follow-up, and an exact one-sided Poisson distribution ($p < 0.05$), the expected number of cases was calculated to be 30.2, and the number of cases needed for statistical significance was calculated to be 40 (Table 4). However, there are only seven observed cases (even after a great deal of media attention and exhaustive search by the manufacturer for all possible cases), less than one-third of the expected cases. Even if the most conservative assumptions are used (from Table 3), the expected number of cases was calculated to be 15.6, which is more than twice the confirmed number of cases of PM/DM.

This postmarket doctor audit survey was conducted by Collagen Corporation to further investigate allegations of an association between implantation of bovine collagen products and the rare autoimmune diseases PM and DM. The results of this study again demonstrate the safety of collagen for clinical use. As an outcome of this research, a number of factors were considered by the FDA for re-evaluating the safety of injectable collagen implants: this postmarket survey, a thorough analysis of the scientific literature by the sponsor, a reanalysis of biocompatibility testing data, and an FDA-sponsored expert review of all available data as part of due diligence efforts by the sponsor. After a review of all information, the FDA responded favorably to a 1994 request to reword the PM/DM warning language for Zyderm and Zyplast. Today, Collagen Corporation continues to safely market Zyderm and Zyplast to many satisfied patients.

4. Conclusions

Current scientific knowledge does not support a biological mechanism for a causal relationship between implantation of bovine collagen (Zyderm or Zyplast) and the onset of PM or DM. In addition, careful evaluation of the duration of follow-up from subject records and the use of reliable background rate of PM/DM from Oddis et al. were used for a more accurate calculation of observed and expected numbers of PM/DM cases. The reported number of observed cases of PM/DM in collagen-treated individuals (seven subjects) is less than one-third of the expected number of cases derived using conservative estimates for the same age, gender, and race (30.2 subjects).

References

1. Pachence, J. M. 1996. Collagen-based devices for soft tissue repair. *J. Biomed. Mater. Res. (Appl. Biomater.)* 33:35–40.
2. Cukier, J., Beauchamp, R. A., Spindler, J. S., Spindler, S., Lorenzo, C., and Trentham, D. E. 1993. Association between bovine collagen dermal implants and a dermatomyositis or a polymyositis-like syndrome. *Ann. Intern. Med.* 118:920–928.

3. Food and Drug Administration. 1991. FDA-convened meeting of scientific experts on collagen and autoimmune diseases. Gaithersburg, MD, October 25.
4. Rosenberg, M. J. and Reichlin, M. 1994. Is there an association between injectable collagen and polymyositis/dermatomyositis? *Arthritis Rheum.* 37:747–753.
5. Lyon, M. G., Bloch, D. A., Hollack, B., and Fries, J. F. 1989. Predisposing factors in polymyositis-dermatomyositis: results of a nationwide survey. *J. Rheumatol.* 16:1218–1224.
6. Oddis, C. B., Conte, C. G., Steen, V. D., and Medsger, T. A., Jr. 1990. Incidence of polymyositis: a 20-year study of hospital diagnosed cases in Allegheny County, PA, 1963–82. *J. Rheumatol.* 17:1329–1334.
7. Hochberg, M.C. 1993. Cosmetic surgical procedures and connective tissue disease: the Cleopatra syndrome revisited. *Ann. Intern. Med.* 118:981–983.
8. Hanke, C. W., Thomas, J. A., Lee, E. W-T., Jolivette, D. M., and Rosenberg, M. J. 1996. Risk assessment of polymyositis/dermatomyositis after treatment with injectable bovine collagen implants. *J. Am. Acad. Dermatol.* 34:450–454.
9. Bohan, A. B. and Peter, J. B. 1975. Parts I and II: polymyositis and dermatomyositis. *N. Engl. J. Med.* 292:344–347; 292:403–407.
10. Stuart, J. M., Townes, A. S., and Kang, A. H. 1985. Type II collagen-induced arthritis. *Ann. NY Acad. Sci.* 460:355–362.
11. Cremer, M. A., Ye, X.J., Terato, K., Owens, S. W., Seyer, J. M., and Kang, A. H. 1994. Type XI collagen-induced arthritis in the Lewis rat. *J. Immunol.* 153:824–832.
12. Kalluri, R., Gattone, V. H., Noelken, M. E., and Hudson, B. G. 1994. The alpha 3 chain of Type IV collagen induces autoimmune Goodpasture syndrome. *Proc. Natl. Acad. Sci. USA* 91:6201–6205.
13. Medsger, T. A., Jr., Dawson, W. N., and Masi, A. T. 1970. The epidemiology of polymyositis. *Am. J. Med.* 48:715–723.
14. Hanissian, A. S., Masi, A. T., Pitner, S. E., Cape, C. C., and Medsger, T. A., Jr. 1982. Polymyositis and dermatomyositis in children: an epidemiologic and clinical comparative analysis. *J. Rheumatol.* 9:390–394.

9

In Vitro Diagnostics
Design of Clinical Studies to Validate Effectiveness

Wayne R. Patterson

1. Introduction

An important category of medical devices, distinct from the therapeutic products described and discussed in other chapters, are the in
vitro Diagnostic Devices (IVDs). IVDs are, by definition of the Food
and Drug Administration (FDA), "...those reagents, instruments
and systems intended for use in the diagnosis of disease, or other conditions, including a determination of the state of health, in order to
cure, mitigate, treat or prevent disease or its sequelae. Such products
are intended for use in the collection, preparation and examination
of specimens taken from the human body."[1] Perhaps more simply
defined, IVDs are laboratory instruments, test kits, or reagent systems.

The FDA has separated medical devices, including IVDs, into three
classes. The specific class designation determines the extent of regulatory control required to ensure safety and effectiveness of the device.[1]
As with many Class I medical devices, Class I IVDs do not require
510(k) premarket notification or conduct of clinical trials. General
controls are sufficient to ensure IVD safety and effectiveness. IVDs
that are classified as Class II devices usually require 510(k) premarket
notification. Clinical trials are required for some IVDs; however, the
size of the clincial trial is dictated by considerations of sample management for the individual IVD product. For example, if an IVD is a
complex system that includes special sample collection procedures,

From: *Clinical Evaluation of Medical Devices: Principles and Case Studies*
Edited by K. B. Witkin Humana Press Inc., Totowa, NJ

extraction of antigens, and a novel detection system, the clinical trial must encompass all phases of the IVD. For Class III IVDs, 510(k) premarket notification is required in some cases and other cases require premarket approval (submission of a Premarket Approval Application or PMA). Clinical trials are required for all Class III devices, because these devices often prevent impairment of human health or present a high risk of injury if the device is misused.[1]

Some IVDs are regulated as biologics and, as such, require product license applications (PLAs) and establishment license applications. IVDs are rarely considered significant risk devices; those that are require an approved Investigational Device Exemption (IDE) for use in clinical research.

2. Guidelines in the Clinical Evaluation of IVDs

In the clinical evaluation and validation of IVD products, the sponsor must consider three distinct and interrelated sets of guidelines: general requirements set forth in the Clinical Laboratory Improvement Act (CLIA); applicable FDA regulations, guidelines, and points to consider; and standards set forth by the National Committee for Clinical Laboratory Standards (NCCLS).

The CLIA provides general requirements for the types of evaluations that laboratories must perform to validate IVDs prior to reporting results on patients. The FDA, as noted previously, has classified all IVDs into one of three medical device classes, all of which are subject to FDA requirements for marketing. For some IVDs, the FDA may specify the "Gold Standard" (assay or method) to which the test assay must be compared, the geographic locations of the sites, the minimum number of samples that must be analyzed, the types of patients from whom to obtain samples, and specific patient demographic information that must be supplied. Additionally, sponsors of clinical studies with IVDs are required to comply with specific FDA regulations regarding product labeling, protection of human subjects, and compliance with good clinical practices.[1] An approved IDE is required by the FDA prior to initiation of clinical studies with significant risk devices (a provision rarely applied to IVDs). Finally, in addition to validation of assay performance characteristics, the FDA requires that Class III IVDs be demonstrated to have "clinical utility." This generally means that the diagnostic information obtained "...contributes to identifying a particular condition or disease," rather than simply characterizing a physiologic parameter.[2]

The NCCLS has established voluntary consensus guidelines for clinical laboratory testing.[3-8] These and other NCCLS guidelines cover administrative as well as technical procedures, and as relates to clinical trials, provide specific evaluation protocol guidelines for laboratories to use as minimum standards for validating laboratory testing systems, reagents, and instruments prior to use for reporting results. They also specify the number of samples required, how to analyze the samples, and how to analyze and interpret the data. Since these are the standards by which all manufacturers' IVDs will eventually be judged, it makes good sense to incorporate these guidelines into clinical trial protocols. In the final analysis, regardless of the exact procedures, methods, or assays used in clinical trials, the requirement for well-designed protocols with specific, defined objectives and endpoints, before beginning the trial, is paramount. Assistance of a biostatistician, preferably with clinical trial experience, is highly recommended to ensure statistically valid protocol design and methodology and later to assist in the analysis and reporting of trial results.

3. IVD Clinical Trials

A brief discussion about clinical trial site selection and clinical investigators is perhaps useful at this point. In all cases, investigators must be qualified to perform the study and must adhere to written protocols in the performance of the evaluation. One useful way to characterize potential investigators relates to the investigator's motive for evaluating IVD products, as follows:

1. "Research scientist": This type of investigator may or may not have experience with clinical trials, but evaluates the product because of a genuine interest in new technology, or an interest in publishing. He or she will usually be compliant with protocol procedures and may even suggest other analyses that would be useful to the evaluation.
2. "Product evaluator": This type of investigator may be considered an opinion leader in the field, and association of the product with the investigator could be perceived as an endorsement of the product. These sites are often approached after the product has received clearance or approval in order to introduce the IVD into a particular geographic location or promote broad general acceptance of the test. Investigators at these sites may consider the clinical trial as an adjunct to their own research and

may not wish to follow a specific protocol. This type of investigator may not be a good choice as a trial site in support of FDA submissions.

3. "Prima donna": This type of investigator may be evaluating several IVDs simultaneously, in some cases as a source of revenue. Most investigators who run this type of business maintain reasonable confidentiality. However, they may consider themselves experts and may deviate from following the written protocol. Disadvantages of this type of investigator are that he or she will move at a slower pace, dictate the terms of the trial, be more expensive, and provide logistical problems for the clinical trial monitor. However, their inclusion may be required so that their name may be associated with the product for marketing purposes after the product has received clearance or approval, or the laboratory community may look to this investigator for favorable comments or endorsement of the product. This may actually be the best site for validating an assay, all things considered, but will almost always take up more time and resources than other sites.

Overlap between types of investigators is expected and no one investigator type is necessarily better or worse than another. Thus, when selecting an investigator and study site, it is most efficient to match the study objectives with the motives of the investigator.

If a clinical trial is not mandatory for an IVD, the question of why to conduct a clinical trial must be considered. There are benefits to conducting the trial at a minimum of one clinical site as a matter of policy, whether required or not. The testing performed in the sponsor's research and development laboratory does not adequately mimic an actual clinical setting. The IVD may perform somewhat differently in a clinical environment because of a mixture of sample types, processing methods, or for other reasons. It is important to know how the product will perform in the hands of the consumer before it is introduced into the marketplace. This practice will provide a high level of confidence at both the manufacturer and user levels and will usually result in favorable publications in scientific journals or presentations at scientific meetings. Also, since laboratories are required to evaluate and validate IVD products before use, participation in a well-designed clinical trial fulfills this obligation. As an added benefit, they may receive payment for the privilege. Laboratory managers may become more interested in your product because of this association, assuming, of course, that the IVD product performs well in their hands during the clinical trial.

4. IVD Investigational Methods—Quantitative Assays

Throughout this section, investigational methods are considered for IVDs, such as clinical laboratory assays, reagents, or systems used to analyze blood or other specimens obtained from human subjects. There are undoubtedly other types of evaluations that will not be discussed. The methods described, if used properly, will provide a good basis for performing IVD clinical studies and thorough evaluation of an IVD's performance characteristics. In the first parts of this section, methods are provided that have proven successful in evaluating laboratory assays submitted for 510(k) premarket notification clearance, and in the later sections, a discussion of similarities and differences in evaluation of IVD products submitted for premarket approval or product license application is presented.

As mentioned previously, one approach to ensuring successful IVD evaluations is to follow NCCLS guidelines to develop valid protocol design and promote recognition and acceptance by clinical laboratories. In general, NCCLS guidelines recommend that for proper evaluation of an IVD product, four steps are required:

1. Each site should review the operation and maintenance of the device, as described by the manufacturer, for an amount of time sufficient to become familiar with all aspects of its use. The amount of time required will vary depending on the complexity of the device.
2. Each site should allot sufficient time to become familiar with the clinical trial protocol. It is recommended that the study monitor review the protocol, in person, with the investigator to ensure complete understanding of expectations.
3. Investigational sites must maintain quality control during the period of the evaluation to ensure proper system operation and function.
4. Sites must collect sufficient data over a period of time in order to allow estimation of long-term performance of the IVD in a clinical setting.[3-9]

The major performance characteristics of IVDs that should be evaluated are precision, sensitivity, linearity, comparison to an FDA approved or cleared IVD, and analysis of an expected range of values for general or specific patient populations. Other characteristics that could be evaluated at clinical sites are specificity or crossreactivity with other analytes,[4] interfering substances,[8] and analyte/matrix interactions and effects. These latter analyses are critical to the overall evaluation of the product, but may be more efficiently performed by the manufacturer's research and development department since commercial laboratories may not have access to potentially crossreacting samples or interfering substances. However, individual investigators may also wish to evaluate these characteristics in their own laboratories.

4.1. Precision

By definition, precision is the ability to consistently obtain the same result on a sample that is repeatedly analyzed. It is customary to report this characteristic as a mean, standard deviation (SD), and percent coefficient of variation (%CV; defined as SD/mean × 100). CV is used extensively by laboratory managers to reflect overall precision performance. The measurement of precision can be further categorized to intra-assay %CV (within-run), interassay %CV (between-run or between-day), and total %CV, which is a combination of both the within- and between-run precision estimates.[3, 10]

A run is defined as any number of samples that are analyzed as a group without stopping the analyzer, and may consist of one or more samples. The test or control samples used to establish precision performance should simulate, as closely as possible, the characteristics of the actual clinical specimens to be analyzed. This is usually accomplished by using commercially available control serum supplied as lyophilized, protein-based pooled human serum that encompasses the entire clinically significant range of analyte concentrations. To ensure that control serum analyte concentrations span all or a significant portion of the systems analytical range, it is best to use bi- or tri-level control serum. These types of control sera are provided as a set of two or three separate vials with distinct analyte concentration ranges for each vial; usually low-normal, mid-range, and high-normal. In addition, if the assay was developed using a particular manufacturers control serum (some manufacturers produce control sera as well as clinical assays), it is prudent to use a different manufacturers control serum for the clinical trial. Laboratories will use a variety of controls and it is often valuable to know and understand how the assay performs with different control sera. The controls should contain analyte concentrations at the low and high ends of the systems analytical range and encompass, as closely as possible, medical decision levels. Every clinical analyte has at least one medical decision level, and ensuring precision at these levels is critical.

A well-designed precision evaluation trial requires enough data to ensure statistically valid estimates of precision that truly reflect the performance of the assay in a clinical setting.[3, 10] NCCLS guidelines specify a requirement for a minimum of 20 "acceptable operating days" in order to obtain sufficient data. The guidelines further describe specific procedures that include the number of runs per day, duplicate analysis of control serum samples per run, changing the order of analysis of the controls, and so forth.[3] It is important to recognize that these guidelines are representative of minimum procedures and that designing the protocol to be more rigorous, especially if it adds to the statistical power, is advantageous. Therefore, as a practice, analyzing each control sample in triplicate for at least two runs per day for at least 20 operating days (this may not be possible for some longer assays) is recommended. This amount of data has been sufficient to provide a good overall estimate of the overall precision that an assay could be expected to provide in a true clinical setting.

It is usually informative and beneficial for study monitors to begin daily analysis of precision data after 3–4 d of operational runs. This helps detect problems with control serum, such as improper reconstitution, mixing, or handling, or possible instrument problems while the evaluation is still in the early stages. From a practical standpoint, it also relieves some of the anxiety surrounding the clinical trial of a new assay when there is pressure to complete the trial quickly and successfully.

The replicate values can easily be set up in a spreadsheet, and the mean, SD, and %CV calculated from the replicates. The question of what constitutes an acceptable %CV is a difficult one to answer. For many assays, a %CV < 10% is considered acceptable, whereas for others, a %CV ≤ 5% is required. Most assays will have a %CV that varies at different analyte levels; it may be important to have a very low %CV only at an analyte level that is near a medical decision level, whereas higher %CV would be acceptable at other analyte levels. As a rule of thumb, the lower the %CV, the better the precision performance. Table 1 represents an example of how a data collection spreadsheet should be designed.

Collection of the data may be performed manually or by downloading electronic data directly into the spreadsheet, which for large studies is significantly more efficient. The within-run, between-run, and total %CVs should be calculated frequently during the trial, accord-

Table 1
Precision Evaluation Sample Data Collection Sheet[a]

Day	Date	Run 1			Run 2			Mean		
		Result 1	Result 2	Result 3	Result 1	Result 2	Result 3	Run 1	Run 2	Daily
1	1/1	160	166	158	157	162	161	161.3	160	160.7
2	1/2	160	164	163	163	159	162	162.3	161.3	161.8
3	1/3	155	153	160	162	154	158	156.0	158.0	157.0
4	1/4	164	160	161	167	160	161	161.7	162.7	162.2

[a]A sample illustration of a partial manual data collection sheet used for recording data utilized in an evaluation designed to estimate assay precision. Even after only 4 d of data collection, a good estimate of assay precision can be made.

ing to NCCLS guidelines[3, 10] and, depending on the proximity of the derived mean to a medical decision level, the %CVs may or may not be acceptable.

4.2. Sensitivity

Sensitivity of a laboratory assay may be reported as either analytical or functional sensitivity. Measurement of each type of sensitivity may or may not be clinically relevant for all assays. Analytical sensitivity is also referred to as Lower Limit of Detection (LLD) and is defined as the lowest result that, with stated probability, can be distinguished from a result of zero when analyzing the zero analyte calibrator or a physiologic sample devoid of the analyte in question.[4] The value obtained for analytical sensitivity is almost invariably lower than the functional sensitivity value, and is frequently used by marketing departments as an advertising point when low analyte levels are critical, and the lowest sensitivity may represent the largest market share for that type of IVD product.

Calculation of analytical sensitivity is as follows: At least 10 replicates from the *same sample cup* containing zero calibrator should be analyzed. The process should then be repeated on two additional sample cups of zero calibrator. The mean and SD of the results are then calculated, reporting the results in luminescent units, optical densities, or other units of measure reported by the system. The analytical sensitivity values must then be determined using the calibration curve, as the value that is 2 SDs above the mean of the zero calibrator results.

Functional sensitivity may be a more clinically relevant measurement for those analytes in which low values are the most important. This measurement is also less difficult to perform, requires less statistical analysis, and does not require determination of the value from the calibration curve. For measurement of functional sensitivity, the technologist serially dilutes a physiologic sample of known analyte concentration to the lowest level that can be consistently detected with %CV less than some target value. Each dilution is then analyzed in duplicate or triplicate. At lower analyte concentrations, the %CV of the measurements becomes greater. The analyte concentration at which the %CV reaches the target value is the functional sensitivity. Figure 1 graphically illustrates how functional sensitivity is determined.

In the illustration, the target %CV is 20% and the analyte concentration at which a 20% CV occurs is 2 U. Therefore, the functional

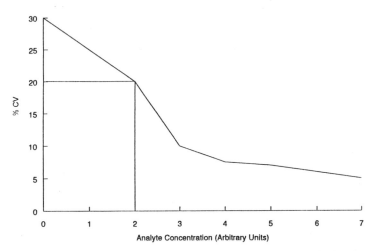

Fig. 1. A plot showing the %CV values after replicate analysis of samples containing various concentrations of an analyte. As the analyte concentration decreases, the %CV of the analyses increases. The analyte concentration at which the %CV of the analyses reaches a predetermined target of 20% is 2 U, which is the functional sensitivity for this particular assay.

sensitivity of this assay is 2 U, and analyte concentrations below a level of 2 U cannot be reliably or consistently measured. Many clinical laboratory managers feel that, for certain analytes, functional sensitivity is the clinically relevant (and therefore, most useful) endpoint. For that reason, it is often measured in addition to (or regardless of) the procedure described in the clinical protocol. Prior knowledge of the assay's functional sensitivity may be useful to the study monitor to prevent being surprised and embarrassed when a shrewd investigator makes a less than flattering discovery.

For many current assays, the advertised stability of the calibration curve is approx 28 d. Therefore, to gain a better insight into the assay, clinical trials often require performance of this type of testing at d 1, 14, and 28 of the calibration curve. This practice can reveal calibration curve stability problems that may ultimately adversely affect assay precision estimates. It is also useful to perform these measurements in the research and development laboratory before going to an external clinical site to prevent unpleasant, potentially embarrassing surprises. Even if a particular measurement is not requested from scientists at an investigational site, the more experienced sites will probably perform the test anyway, especially for those assays in which analytical sensitivity is crucial.

4.3. Linearity

Linearity is defined as "...the measure of the degree to which a curve approximates a straight line."[5] It must be recognized that linearity is a characteristic of an analytical system and is distinct from accuracy and precision. To a clinical laboratory, it is another measurement of overall system performance. To manufacturers, measurement of linearity allows the establishment of claims for an assay. This measurement may also be referred to as "dilution recovery" for reasons that will become obvious.

There are many different ways to approach the measurement of assay linearity and investigational sites may have, and be insistent on using, their own method. The "in-house" method should be evaluated regarding its ability to measure system linearity over a wide range of analyte concentrations, and if it accomplishes the desired objective, the preferred method at that site may be used.

One method that has proven successful is to analyze a sample or sample pool with a known analyte concentration that has been diluted to at least five different analyte concentrations. NCCLS recommends that the original sample or sample pool have an analyte concentration up to 30% higher than the upper limit of the analytical range of the assay.[5] The vehicle for dilution must provide a matrix that is compatible with the assay and system. For example, many assays require a protein-based matrix to assure optimum system performance. In these cases, using a diluent, such as saline or distilled water, would result in inaccurate data and, therefore, invalidate the analysis. The sample should instead be diluted using another patient sample with a low analyte concentration of perhaps the "zero" analyte calibrator used to calibrate the system. The dilutions are then assayed in quadruplicate for the analyte in question and the data plotted. Figure 2 illustrates a plot of data collected from a high analyte concentration sample diluted with another patient sample and assayed at five different analyte concentrations. The evaluation of linearity consists of fitting a straight line to the data by the least squares method.[5, 11] Visual inspection of the plot may be sufficient to determine system linearity, requiring no further statistical analysis. However, if nonlinearity is apparent or if there are obvious outliers, more in-depth statistical analysis is indicated, as described elsewhere in the literature.[5]

In other instances, it may be desirable to evaluate linearity over an analyte concentration range from zero to some higher value in order

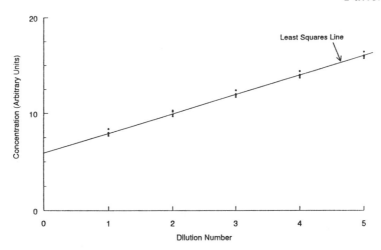

Fig. 2. A plot designed to visualize and determine linearity for a laboratory assay. Values were obtained by diluting a patient sample to various
concentrations within the analytical range of the assay and then measuring
the analyte concentration in each dilution. Visual inspection would confirm
that the assay is linear over the concentration range tested.

to make a specific claim. In this case, the patient sample with high
analyte levels should be diluted with the zero analyte calibrator supplied by the manufacturer to at least five analyte concentrations. The
sample dilutions as well as the zero calibrator are then assayed in
quadruplicate and the assayed (observed) concentrations should be
plotted vs the calculated (expected) concentrations. Alternatively,
commercial standards of known concentration may be analyzed and
the data plotted, assuming that the stated concentration of the commercial standards is correct. It is recommended that the assayed concentrations be within 10% of the expected concentrations, which
should take into account pipeting or other technical errors associated
with dilution of the samples. Figure 3 illustrates a plot in which the
laboratory or the manufacturer evaluates linearity over the entire
assay range. The line on the plot represents the fitted curve from least
squares analysis[11] and should approximate a straight line if the system
is linear.

Calculations of linearity are sometimes very revealing. If the curve
becomes nonlinear at higher analyte concentrations (*see* Fig. 3), it
could mean that there were calibration problems. Also, if the observed
and expected concentrations are not within 10% of each other, it

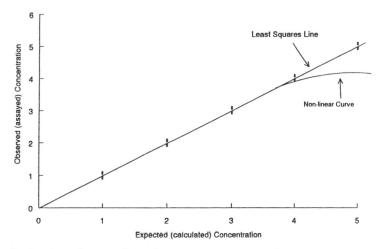

Fig. 3. A plot of assay linearity over a concentration range from 0 to a value of 5 U. If least squares analysis reveals a straight line, the assay is linear. The nonlinear section of the assay between 4 and 5 U could indicate a system problem. In any event, the reason for nonlinearity must be investigated unless this phenomenon is expected or described and explained by the manufacturer. If this is the case, the true analytical range that should be used is likely from 0 to 4 U.

could indicate that the reagents used in the dilution result in high background noise, disrupting the assay. In this case, the diluent may be interfering with the assay results.

4.4. Method Comparison

This type of experiment is designed "...to evaluate the bias between two assays (methods) which measure the same analyte."[6, 12] In the method comparison evaluation, a group of prospective and/or retrospective samples are analyzed with both the test method and a reference (or comparative) method. The reference method may be the current method used by the laboratory, the method that was used by the manufacturer to establish claims for the assay, or another widely recognized and accepted method. Using the appropriate experimental design, an estimate can be made of the difference between the two methods. NCCLS guidelines recommend using at least 40 patient samples analyzed in duplicate, with the two methods conducted within 2 h of each other, over a 5-d period.[6] Confidence in the test assay is expected to increase with a greater number of patient samples. NCCLS also recommends that at least 50% of the samples analyzed should contain analyte concentrations outside the expected reference range. Analysis may be performed on fresh or stored specimens, providing that storage conditions are sufficient to guarantee sample stability and are in accordance with accepted policies and procedures. For some assays, obtaining specimens with low or high analyte concentrations or from special patient populations may be difficult, and it may be necessary to "bank" these special samples as they are obtained. Most investigators experienced in IVD clinical trials have sample banks from which they can pull unique and interesting specimens to see how the test assay will perform with potentially troublesome samples. These same investigators will likely be quick to point out instances when the test assay does not perform as anticipated and may offer potential explanations for the discrepancies. Manufacturers and investigators must recognize that this type of evaluation does not identify specific sources of bias for either method, and differences between methods could result from many contributing factors.

In order to compare the test and reference methods, the mean of duplicate values for each sample is then plotted. For each sample, the value obtained with the test assay is customarily plotted on the Y-axis and the value from the reference assay is plotted on the X-axis. Regression analysis is performed and the slope of the regression line, the Y-intercept, and correlation coefficient are calculated.[6, 12] When performing regression analysis, the tendency is to assume that the reference assay (X-axis) is without error. This is rarely the case with clinical laboratory assays. However, by assaying samples with wide-ranging analyte concentrations, the error effects are usually quite small and have little effect on regression estimates. The NCCLS guidelines[6] review the calculations and recommend that r values should be ≥ 0.975 or $r^2 \geq 0.95$ in order to adequately calculate the slope and Y-intercept. If the r values are below the specified levels, more samples must be analyzed. From a practical standpoint, additional samples may not be required if the specificity of the assays differ. In this instance, the r value may be significantly below 0.975, but the variability lies with the comparative (reference) assay and the test assay may actually be the better assay. It may be necessary, in some instances, to truncate data at points where nonlinearity occurs, especially if it occurs at the extremes of the analytical range. This will only be appropriate if the data points removed are not within the clinically significant range. If no linearity between the two methods is present, NCCLS specifies

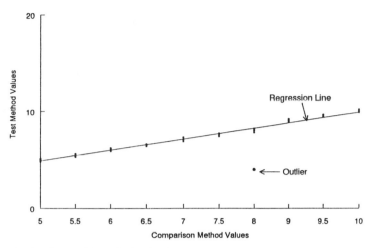

Fig. 4. A plot, with relatively few sample data points, derived from a method comparison study that illustrates good correlation between the two methods, as well as one outlier. The outlier sample should be investigated to determine reasons for lack of correlation.

that the linearity of each assay must be verified independently. In deference to NCCLS guidelines, it is desirable to use a minimum of 150 samples for 510(k) assays. The required number may be higher depending on the assay and FDA requirements for the specific assay or category of assay. Figure 4 demonstrates the principal of method comparison for very few samples and is meant for illustration only.

After plotting the data, visually check the plot for obvious outliers. NCCLS provides guidance for determining if a data point is an outlier and allows removal of a data point if certain criteria are met.[6] Whether a data point is removed or not, it is critical to understand why the data point is an outlier. The analysis should be repeated by both assays, and if the values are confirmed, an in-depth investigation of the sample condition, storage history, interfering substances, patient history, and other possible variables should be undertaken. To the extent possible, the reason for nonlinearity must be explained. A thorough understanding of patient or sample variables that adversely effect assay results is critical to the overall success and acceptance of the assay in clinical laboratories.

Questions will invariably arise regarding what constitutes an acceptable slope, intercept, and correlation coefficient. The answer depends on the test assay, the reference assay, and other variables associated with the comparative evaluation. A high *r* value may be critical for one assay but a Y-intercept close to zero may more important for another assay. Manufacturers may develop their assays to be in direct competition with another manufacturer's assay, and in these cases, one should attempt and expect to obtain a slope of the line of regression that is very close to 1.0, a Y-intercept very close to zero, and *r* values >0.95. For assays considered to be "front runners" (vast improvements over current technology), attempts to achieve the above values may not be appropriate, and marketability of the assay will depend on precision, improved technology, or other features. As with other performance characteristics of the assay, knowledge of the requirements and expectations of the laboratory as well as the characteristics of assays manufactured by competitors form the basis for the expectations of the new technology.

4.5. Assay Bias

Another type of data analysis that can be useful and informative is measurement and graphing of assay bias. NCCLS guidelines[6] give examples and illustrate the computations, that for purposes of brevity, are not discussed in detail here. The bias of an assay is determined by measuring the difference between a given data point and the regression line for each data point and then calculating the standard error of the estimate of the differences, an estimate of predicted bias at a given medical decision level, and a 95% confidence interval for the true bias. Many clinical laboratories prefer to visualize assay bias on a graph, and it is also useful for manufacturers to be aware if the assay is biased in either the positive or negative direction. Figure 5 illustrates how one might graphically represent the data from a method comparison study in the form of a bias plot.

In this figure, the mean of the comparison method is plotted as the actual value on the X-axis vs the difference between the mean of the test and comparison methods, as a positive or negative number, on the Y-axis. For instance, if the mean of the comparison method replicates was 75 and the mean of the test method replicates for the same sample was 70, the data would be plotted as 75 on the X-axis and -5 on the Y-axis. This method allows a rapid visual picture of how the test assay responds at different analyte concentrations vs the comparative assay and may show a bias, positively or negatively, at different

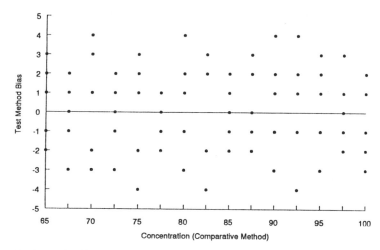

Fig. 5. A method comparison bias plot illustrating the difference in values between the comparative and test methods. This plot shows a relatively even distribution and close correlation of the test method with the comparative method.

analyte levels. Also, it may illustrate that, if the Y-axis data points are consistently above or below the zero line, the test assay is probably biased in a particular direction when compared to a particular assay. This is not necessarily good or bad and will need to be interpreted by the manufacturer. The manufacturer may desire bias in a certain direction for a particular assay.

4.6. Expected Range

An expected range is the range of analyte concentrations that would be consistent with a disease-free state in patients.[7, 13, 14] All laboratories are required to establish expected ranges for each assay they perform, which is typically accomplished by analyzing samples from a group of patients that, using an appropriate and acceptable screening procedure, are verified as disease-free. In theory, if the slope of the regression line obtained in comparing the test assay with the reference assay is 1.0, then the expected range of the test assay would be identical to that of the reference assay and there would be no need to perform the analyses for the test assay. However, this is not always the case and expected range values should always be established by the laboratory prior to implementing the assay. Expected ranges for analytes can and are affected by many variables, including, but not limited to, geographic location, ethnicity of the patient population, sex of the patients, and other factors. In the past, many laboratories determined their expected ranges by simply using those ranges published by the manufacturer in the product insert. This is a potentially dangerous practice since the expected range of many analyte concentrations are known to vary according to the ages or sex of the patients, geographic location, patient ethnicity, and diet. The expected range may or may not reflect a healthy state, but may be normal for a specific population or geographic area. How the values are interpreted is a matter for the patient's physician insofar as the physician is aware of the expected range. Once established, the expected range values should not be allowed to become static. Expected ranges should be checked periodically and readjusted if necessary, especially if there are major changes in population demographics.

5. IVD Investigational Methods—Qualitative Assays

Although the type of data obtained from qualitative assays differs significantly from that derived with quantitative assays some of the same types of evaluative analyses can (and probably should) be performed on qualitative assays. Generally, qualitative assays give results that are interpreted as either positive or negative but meaningful data on assay effectiveness can and must be obtained. For qualitative assays, data can be generated and extracted from which estimates of precision, sensitivity, specificity, and positive and negative predictive value may be made. The following sections provide a brief description of how these types of evaluations may be performed.

5.1. Precision

For assays that report results as either positive or negative, precision estimates are a bit more of a challenge, but it is just as important to validate reproducibility in these assays as it is with quantitative assays. Qualitative assays may generate data in the form of luminescent, optical density, or other measurement units from which a positive or negative result is derived based on a predetermined range or cut-off value. Although the end result is not amenable to mathematical manipulation, the measurement units used to derive that result are numerical and capable of undergoing statistical evaluation. It is possible to calculate a mean, SD, and %CV on the actual measurement units that, if the assays were performed using an automated laboratory analyzer, may have to be extracted from computer archived files. The %CV calculated from these units may be higher than the norm for quantitative assays, so confidence in the assay may be slower in attaining. For instance, it is not uncommon to obtain %CVs in the range of 20–30% for negative control samples. Positive control sample %CV's should be <20%. One reason for the higher %CVs is that calibration of these assays is performed with each run so the range of measurement unit values obtained will usually be larger. The end result, either positive or negative, should always be the same, but the actual measurement unit values used to calculate precision will vary within a pre-established range.

5.2. Sensitivity

The definition of sensitivity (the ability of the assay to detect an analyte at very low concentrations) is the same whether or not a quantitative or qualitative assay is being evaluated; however, the procedures are somewhat different. With qualitative assays, the investigator is evaluating the assays ability to detect an analyte at low levels in comparison to a "Gold Standard" assay specified by the FDA and assumed to be correct. A problem arises if the "Gold Standard" is outdated and the assay being tested is actually more sensitive than the specified standard. In these cases, investigators should work with the FDA to identify some other method to rectify this discrepancy. Figure 6 illustrates how data may be arranged to calculate sensitivity.

In this illustration, the test method obtained 2 negative and 249 positive results on samples in which the comparative method obtained 251 positive results. The sensitivity is then calculated as: $(249/251) \times 100 = 99.2\%$. A discussion regarding the relationship between sensitivity and assay efficacy is included in Section 5.4.

5.3. Specificity

Specificity determines the ability of the test assay to correctly identify samples containing the analyte and not incorrectly identify negative samples as positive. Figure 7 illustrates how the data from this type of evaluation might be graphically represented.

In this illustration, the test assay obtained 302 negative results and no positive results on a total of 302 samples. The comparative assay recorded the same results and the specificity is calculated by the formula: $(302/302) \times 100 = 100\%$. A discussion of the clinical significance of this result is found in the next section.

Comparative Method

	Negative	Positive
Negative	0	2
Positive	0	249

Test Method

Fig. 6. A graphic representation of the results of qualitative testing of 251 known positive samples with a comparative and test method. In this case, the test method obtained negative results on two of the known positive samples.

Comparative Method

	Negative	Positive
Negative	302	0
Positive	0	0

Test Method

Fig. 7. A graphic representation of the results of qualitative testing of 302 known negative samples with a comparative and test method. In this case, both the test and comparative methods agreed on all samples.

Comparative Method

	Negative	Positive
Negative	302	2
Positive	0	249

Test Method

Fig. 8. A graphic representation of all samples qualitatively tested with a comparative and test method. The data result from a combination of the data displayed in Figs. 6 and 7. PPV and NPV are determined from these data.

5.4. Negative and Positive Predictive Values

The negative predictive value (NPV) is, in essence, the probability that a negative result is obtained on a sample that contains none of the analyte in question. Perhaps more simply stated, it is the probability that a sample testing negative for the analyte really is negative for the analyte. Conversely, the positive predictive value (PPV) is the probability that a positive result correctly identifies a sample containing the analyte or a positive really is positive. Figure 8 shows an example of how the data may be arranged graphically to illustrate NPV and PPV. In this figure, the data reveal that the test method obtained 302 negative results from 304 samples devoid of the analyte and the calculation of NPV is performed using the formula: (302/304) × 100 = 99.3%. The same figure shows that the test method obtained 249 positive results out of 249 samples containing the analyte and the calculation of PPV is performed using the formula: (249/249) × 100 = 100%.

The significance of these values to a manufacturer or clinical laboratory manager may be questioned. The answer depends on the specific assay and the customer's requirements (perceived or real) for the assay. For at least some qualitative assays under development, a balance must be achieved between sensitivity and specificity. In other words, in order to get increased sensitivity, there may have to be some sacrifice in assay specificity. For some assays, the assay sensitivity is critical to rapid and efficient treatment of a patient, and occasional false positive results would be acceptable.

For another assay, it might be more important to have a more accurate result (greater specificity) and the laboratory manager would be willing to sacrifice some sensitivity in order to maximize specificity. In this case, an occasional false negative result may be acceptable. In a general sense, the NPV relates to sensitivity and the PPV relates to specificity and both give some sense of confidence regarding the believability of the results from the assay. Based on input from the end-users (customers), assays are designed with certain performance characteristics in mind. Investigators and clinical trial monitors must be familiar with the assay specifications and evaluate the significance of the data with the assay specifications in mind.

6. Class III IVD Trials

The methods described thus far have proven successful in the past for submission of laboratory assays for 510(k) clearance. There are some similarities and differences in design and implementation of clinical studies of IVD products that require premarket approval or product license applications. Generally, the requirements for PMAs or PLAs are significantly more stringent and demanding because manufacturers must prove product safety as well as effectiveness.[15] To this end, these types of assay evaluations will require significantly more sample analyses and, perhaps, data on other clinical endpoints. In designing the studies, one should refer to specific FDA guidelines or points to consider to help determine the number of samples and the types of patients from whom the samples are to be obtained. For many assays, especially if specificity is critical, the FDA will require that samples be obtained from patients with similar disease states or symptoms in order to determine the extent of crossreactivity with other analytes. In these assays, with patient safety in mind, it may be deemed more important to show specificity than to provide sensitivity to the analyte at extremely low concentrations. Another consideration may include the requirement for geographically diverse clinical trial sites. This requirement is intended to assure that sensitivity, specificity, and other assay characteristics are similar in different populations and geographic areas; for example, East vs West coast or urban vs rural populations. FDA guidelines may also include a requirement for determining crossreactivity or interference by similar analytes. Some of the requirements may seem at first to be excessive, but one should remember that the intention is to ensure safety as well as effectiveness, and the additional samples and sample types are intended for this purpose.

Two examples of IVD products for which PMAs have been submitted are assays for tumor markers, which are currently being considered for reclassification as Class II devices, and α-fetoprotein assays. Examples of IVD products for which product license application assays that would require stringent and lengthy clinical trials are assays for hepatitis A, B, or C; human immunodeficiency virus (HIV); and human T-cell leukemia virus (HTLV). The risk of harm (physical and psychological) to the patient is significant, if the patient is misdiagnosed or if the assay is misused. Therefore, the safety and effectiveness of these assays are critical, and collection of extensive clinical data is imperative.

7. Conclusion

The specific types of evaluations required for clinical trials for IVDs in support of submissions for premarket notification and clearance, premarket approval, or PLA do not differ significantly. However, there may be specific differences among the requirements for the minimum number of samples in the assays, number and geographic location of trial sites, assurance of population diversity, the requirement for specific patient populations, and the requirement for an IDE.

There may also be differing opinions regarding what is expected from a clinical study for an IVD. Clinical and regulatory departments are generally interested in ascertaining that the product demonstrates "substantial equivalence" to an FDA-cleared product, or is as safe and effective as an FDA-approved product, and that data show adequate purity and potency for licensed products. At the same time, a well-designed and implemented clinical investigation can meet the needs and goals of other departments within the company. Demonstration that the assay compares as well as (or perhaps better than) an assay currently in the marketplace is also valuable to the ultimate success of the assay. Clinical evaluations of IVD products, even if not required by the FDA, are of critical importance. A well-designed study with appropriate and extensive statistical analysis can reveal not only the performance characteristics of a laboratory assay, but also its critical limitations. Significant time should be invested in these processes before the product is even taken to a clinical study site. The sponsor is hopeful that most of the knowledge gained will be positive, but even negative information can be valuable to the overall understanding of the product, the resolution of issues that may arise surrounding the assay, or avoidance of potential performance issues associated with registration prior to marketing. The information contained in this chapter will be useful to those individuals embarking on new clinical studies, or perhaps to those individuals with clinical trial experience.

References

1. Food and Drug Administration, Health and Human Services. 1995. *Medical Devices.* 21 CFR §800–812.
2. Food and Drug Administration, Center for Devices and Radiological Health, Office of Device Evaluation. 1991. *Clinical Utility and Pre-Market Approval.* PMA Memorandum, May 3, p. 1.
3. National Committee for Clinical Laboratory Standards. 1992. *Evaluation of Precision Performance of Clinical Chemistry Devices,* 2nd ed. NCCLS Document EPT-T2, Vol. 12, No. 4, NCCLS, Villanova, PA.
4. National Committee for Clinical Laboratory Standards. 1987. *Assessment of Clinical Sensitivity and Specificity of Laboratory Tests.* NCCLS Document GP10-P, Vol. 7, No. 6, NCCLS, Villanova, PA.
5. National Committee for Clinical Laboratory Standards. 1986. *Evaluation of Linearity of Quantitative Analytical Methods.* NCCLS Document EP6-P, Vol. 6, No. 18, NCCLS, Villanova, PA.
6. National Committee for Clinical Laboratory Standards. 1993. *Method Comparison and Bias Estimation Using Patient Samples.* NCCLS Document EP9-T, Vol. 13, No. 4, NCCLS, Villanova, PA.
7. National Committee for Clinical Laboratory Standards. 1992. *How to Define, Determine, and Utilize Reference Intervals in the Clinical Laboratory.* NCCLS Document C28-P, Vol. 12, No. 2, NCCLS, Villanova, PA.

8. National Committee for Clinical Laboratory Standards. 1986. *Interference Testing in Clinical Chemistry*. NCCLS Document EP7-P, NCCLS, Villanova, PA.

9. Peters, T. and Westgard, J. O. 1987. Evaluation of methods. In *Fundamentals of Clinical Chemistry* (Tietz N. W. ed.). 3rd Ed., Saunders, Philadelphia, pp. 225-237.

10. Carey, R. N. and Garber, C. C. 1984. Evaluation of methods. In *Clinical Chemistry: Theory, Analysis, and Correlation* (Kaplan L. A. and Pesce A. J. eds.). C.V. Mosby, St. Louis, pp. 338-359.

11. Cornbleet, P. J. and Gochman, N. 1979. Incorrect least-squares regression coefficients in method comparison analysis. *Clin. Chem.* 25:432-438.

12. Bookbinder, M. J. and Panosian, K. J. 1987. Using the coefficient of correlation in method comparison studies. *Clin. Chem.* 33:1170-1176.

13. Brown, G. W. 1984. What makes a reference range? *Diag. Med.* Jan.: 61-69.

14. Garber, C. C. and Carey, R. N. 1984. Laboratory statistics. In *Clinical Chemistry: Theory, Analysis, and Correlation* (Kaplan L. A. and Pesce A. J. eds.). C.V. Mosby, St. Louis, pp. 287 300.

15. Food and Drug Administration. 1993. *The Pre-Market Approval Manual*, FDA #93-4214. U.S. Government Printing Office, Washington, DC.

10

A Controlled Study of Intra-Articular Hyalgan® in the Treatment of Osteoarthritis of the Knee

Roberto Fiorentini, Frank C. Dorsey,
Sharon A. Segal, Roland Moskowitz

1. Introduction

A US multicenter clinical trial was undertaken to evaluate the safety and effectiveness of Hyalgan®, a defined molecular weight fraction of highly purified hyaluronic acid (500–730 kDa), prepared as a buffered solution in physiologic saline. Hyalgan is injected intra-articularly for the sustained relief of pain and joint dysfunction in osteoarthritis of the knee. This product is classified as a device according to the current ruling of the US Food and Drug Administration (FDA). However, this categorization may be questionable. Other regulatory authorities consider hyaluronic acid products as drugs because the mechanism of action appears to go beyond the mere physical function of replacing abnormal synovial fluid ("viscosupplementation").

Four design issues required resolution for the successful completion of this clinical trial. First, since the primary endpoint of effectiveness is inherently subjective (reduction in pain), it was recognized that it was important to employ the most reproducible and quantitative measures of pain and joint function available and to maintain the strictest possible double-blind conditions. The next issue involved a problem common to all clinical trials on medical devices. The skill and experience of the implanting physician/investigator are primary

From: *Clinical Evaluation of Medical Devices: Principles and Case Studies*
Edited by K. B. Witkin Humana Press Inc., Totowa, NJ

determinants of the success of an implanted device; these are relevant to the Hyalgan multicenter clinical trial because the intra-articular injection technique and reporting were expected to vary between sites and investigators. A third design issue concerned identifying appropriate control groups. In order to demonstrate the clinical utility of the test device, both a placebo and a standard treatment group were used as controls. However, in a long-term study of osteoarthritis, it is not ethical to include a pure placebo group without providing an escape analgesic. It was determined that the only acceptable standard treatment was an oral nonsteroidal anti-inflammatory drug (NSAID). The final design issue was that the criteria for success had to be carefully and specifically defined for this clinical trial as a regulatory requirement to demonstrate the clinical utility of the device. This chapter describes the design, conduct, and results of the Hyalgan multicenter clinical trial, including the resolution of the issues listed above.

2. Clinical Trial Objectives

The primary objectives of this clinical trial were to determine whether a cycle of intra-articular injections of a hyaluronic acid product in patients with osteoarthritis of the knee is safe and more effective in relieving pain and joint dysfunction than injections of vehicle (placebo) with up to 4 g acetaminophen per day as an escape analgesic, and how the test product compares to a commonly prescribed NSAID (naproxen).

3. Clinical Trial Design

This clinical trial was a double-blind (masked observer), double-dummy, placebo and NSAID controlled, multicenter prospective clinical trial. A total of 495 subjects with moderate to severe pain were randomized into three equal-sized groups: test group (Hyalgan), placebo group (vehicle), and NSAID group (naproxen). Each group's specific treatment is described below.

Subjects in the test group (Hyalgan) received 1% lidocaine injected subcutaneously (sc) into the knee followed by intra-articular injection of 20 mg/2 mL Hyalgan; synovial fluid (when present) was aspirated. This procedure was repeated once per week for a total of five injections. Subjects also received dummy capsules for placebo naproxen to be taken orally twice per day for 26 wk.

Subjects in the placebo group received 1% lidocaine sc into the knee followed by an intra-articular injection of 2 mL phosphate-buffered saline (PBS); synovial fluid (when present) was aspirated. This procedure was repeated once per week for a total of five injections. Subjects also received dummy capsules for placebo naproxen to be taken orally twice per day for 26 wk.

Subjects in the NSAID group (naproxen) received 1% lidocaine sc in the knee (sham injection); synovial fluid was not routinely aspirated. This procedure was repeated once per week for a total of five sham injections. Subjects also received 500-mg tablets of naproxen to be taken orally twice per day for 26 wk.

All subjects received 500-mg acetaminophen tablets to be taken as needed, only when absolutely necessary, and not to exceed 4 g (8 tablets) per day. Use of all other pain relievers and/or anti-inflammatory medication was precluded by the study protocol.

Special care was taken to ensure double-blind conditions. Since the viscosity of the product makes it easily distinguishable from the vehicle (placebo), all effectiveness evaluations were performed by a *masked observer* who did not witness the injection procedure. The masked observer was a physician, research nurse, or metrologist certified by the principal investigator to make clinical measurements in this study. (A masked observer who was a nonphysician was required to consult a study physician to certify a subject's initial eligibility.) For each subject, all evaluations were to be performed by the same masked observer throughout the study.

The injection-procedure physician administered the injection and was therefore unblinded to the treatment the subject received. This physician prepared and draped the subject's knee so that the subject was unable to observe the procedure. This same physician *always* administered the injection but did *not* evaluate the clinical status of the study knee. The injection-procedure physician recorded all adverse events and concomitant medications. The duration of the procedure was the same, whether an actual injection or a sham injection was performed.

4. Schedule of Clinical Trial Procedures

After meeting initial screening requirements, all subjects were removed from any NSAID therapy for a period of 2 wk and were allowed only acetaminophen as needed for pain relief. After 2 wk, all subjects returned for baseline evaluations. Subjects who were eligible to participate in the clinical trial were stratified within center on the basis of moderate vs marked pain, as determined by the masked observer at the baseline evaluation.

All subjects who tolerated the NSAID washout and met all entry requirements received their first injection immediately after stratification/randomization. Intra-articular injections (test, placebo, or NSAID sham) were administered weekly for a total of five injections (wk 0–4). Subsequent visits and evaluations took place at 5, 9, 12, 16, 21, and 26 wk.

5. Clinical Trial Population

Six hundred-seven patients were screened at 15 participating centers for inclusion in the clinical trial. After screening, the enrolled population consisted of 495 patients (206 males and 289 females). Patients met the following inclusion criteria: diagnosis of idiopathic osteoarthritis of the knee (by American College of Rheumatology criteria), including patients with grades 2 or 3 osteoarthritis by Kellgren's criteria, and moderate to marked pain as assessed by the masked observer.

A subject was considered ineligible for enrollment into the study if he or she met any of the following exclusion criteria: primary inflammation of the knee; inability to perform the 50-foot walk test; grade 0, 1, or 4 osteoarthritis by Kellgren's criteria; large axial (mediolateral) deviations; peripheral neuropathy; and poor general health.

6. Success Criterion

Based on discussions with the FDA, treatment success for the purpose of this clinical trial was defined *a priori* as a reduction in pain determined by a combination of positive results on four primary effectiveness measures to ensure that a successful outcome would represent a clinically significant benefit.

6.1. Primary Effectiveness Measures

1. Visual Analog Scale (VAS) for pain: The primary effectiveness measure was the VAS for pain during the 50-foot-walk test. VAS is a sensitive, validated, and widely used method for assessing pain. The scale provides the subject with a robust and reproducible method of assessing pain severity. The method has demonstrated a good correlation with verbal

rating scales, excellent sensitivity to change during interventional trials,[1] and a high degree of reproducibility on repeated pain measurements.[2] Briefly, a single line (commonly 100 mm long) is drawn to represent the continuum of pain. Extremes are labeled as "no pain" and "as severe as it could be." Subjects mark the line at the point corresponding to their estimate of pain, and the distance from zero to the mark represents the severity of pain. Differences between baseline measurements and subsequent pain estimates may be calculated over time in clinical trials.

A positive outcome for this variable was defined to be a significantly greater reduction on the VAS for test group subjects when compared to placebo-treated subjects ($p \geq 0.05$). Such a difference was also to exceed one-fourth of a standard deviation (SD) of the change from baseline on the VAS.

2. Categorical assessment of pain: Analyses of the categorical assessment of pain (0, none to 5, disabled) for 48 h preceding visits by both the subject and the masked observer were to be concordant with the VAS results for the clinical trial to be considered successful. That is, the effect on pain was to change in the same direction using both criteria, but the changes were not necessarily statistically significant.

3. Magnitude of effects in test vs placebo groups: The magnitude of the observed effects for test subjects vs placebo subjects on both the VAS and the categorical pain assessment scale were to be at least 50% of those observed for the NSAID group vs placebo.

4. Maximum amount of severe pain and swelling: In order for the trial to be considered successful with regard to safety, it was decided *a priori* that the incidence of severe pain and swelling consequent to intra-articular injection of the test product in this study should occur < 5% of the time for the device to be considered clinically useful.

6.2. Secondary Effectiveness Measures

Secondary effectiveness measures were as follows: the Western Ontario and McMaster University (WOMAC) Osteoarthritis Index, the time to perform the 50-foot-walk test, measurement of the knee range of motion (extension and flexion), measurement of heel-to-buttock distance, effusion (patellar ballottement, bulge sign, and synovial fluid aspiration if performed), knee circumference, acetaminophen consumption, and the global assessment of treatment effectiveness by the subject and a masked observer.

6.3. Safety

The safety of test products was evaluated by a number of parameters. Routine clinical laboratory tests included hematology, serum

chemistries, and urinalysis. Synovial fluid analysis was also conducted. Adverse events were recorded at each visit.

7. Statistical Methods

The clinical decision to elect intra-articular injections over other
therapies must be based on the likelihood of long-term success; therefore, the principal endpoint of this trial was chosen to be the 6-mo
result in completers (i.e., safety and effectiveness at 6 mo for those
patients who remained in the trial until then). A previous clinical trial[3]
indicated that the expected effects of the test product (change from
baseline) compared to placebo at 6 mo were large enough (one-third
to one-half SD) that 120 patients per treatment arm would provide
> 80% power for analyses of either categorical outcomes or the VAS
for pain on the 50-foot-walk test. It is interesting to note that, had an
intent-to-treat design been used, the attenuation of effects by incomplete treatment and by increased variability would have required
exposing roughly four times as many patients to the experimental
therapy to yield the same power.

Different effect sizes were anticipated for the various centers and
for the two severity strata. Therefore, continuous variables were analyzed using an analysis of covariance (ANCOVA) model with main
effects for treatment, center, and stratum using the baseline value as
a covariate. The main effects for treatment were evaluated by two-
tailed tests; and treatment by center and treatment by stratum interactions were tested separately and considered significant if $p \le 0.10$.
Two-tailed Fisher's Exact Tests were used for ordinal categorical
variables. The principal effectiveness analyses as defined *a priori* required a significant ($\alpha = 0.05$) ANCOVA comparison of the test
product to placebo with respect to the VAS for pain on the 50-foot-
walk test at wk 26, concordant results on the categorical outcomes,
and effect sizes of the test product relative to placebo at least one-
half as large as those for the NSAID group compared to placebo.

In addition, the same ANCOVA model was applied to all three
treatments in order to obtain comparable model-adjusted means. Only
the *p*-values in the principal analyses should be considered as measuring statistical significance. Because of the multiplicity of tests, some
of which were requested *post-hoc* by the FDA, other *p*-values should
be considered only as relative measures of difference. For safety anal-

yses, in order to follow a generally accepted conservative convention, *p*-values ≤ 0.10 were considered statistically significant.

8. Clinical Trial Results

8.1. Demographics and Baseline Characteristics

The demographics of clinical trial participants were comparable across treatment groups with regard to age, sex, race, height, weight, medically relevant characteristics and abnormalities, history of osteoarthritis, prior NSAID use, physiotherapy history, and weight bearing/use of assistive devices. The mean age of the 495 study participants was 63.7 ± 9.8 yr (range 40–90 yr); 58.4% of the participants were female and 41.6% were male; 81.8% were Caucasian, 16.2% were African-American, and 2.0% were classified as "other"; the mean height was 168 ± 10.5 cm (range 102–198 cm); and the mean weight was 88.7 ± 18.2 kg (range 45–170 kg).

8.2. Discontinuations

Of the 495 subjects enrolled in the clinical trial, 162 discontinued prior to study completion, resulting in a final study cohort of 333 subjects (test, 105; placebo, 110; and NSAID, 118) who completed the entire 26 wk. The number of subjects who discontinued was comparable across treatment groups (test = 59, placebo = 53, and NSAID = 50). Demographic and medical characteristics of the 333 completers were comparable to those of the 495 subjects enrolled. The major reasons for discontinuation were adverse gastrointestinal events, loss to follow-up, lack of effectiveness, other medical problems, and other musculoskeletal pain.

8.3. Effectiveness

The statistical analyses for both primary and secondary effectiveness measures for completers at wk 26 are summarized in Table 1. As specified in the success criterion, the primary effectiveness analysis (requiring a statistically significant difference) pertains to the comparison between the test group and placebo on the VAS at wk 26. Summary descriptive statistics for all three treatments are displayed in Table 1 to permit evaluation of the clinical utility of the test product relative to placebo and NSAID. The *p*-values of the secondary analyses presented should be considered as relative measures of effect sizes for comparative purposes only.

Table 1
Summary Table of Primary and Secondary Effectiveness Endpoints for All Completers at 26 Wk

Measure	Test vs placebo		Test vs NSAID		Test vs placebo vs NSAID
	Effect size (SE)[a]	p-value	Effect size (SE)	p-value	p-value
VAS for pain on 50-foot-walk test (mm)	8.85 (3.07)	0.004	4.12 (3.14)	0.191	0.032
Time for 50-foot-walk test (seconds)	0.55 (0.52)	0.285	0.14 (0.46)	0.769	0.563
WOMAC-A (pain)	5.56 (2.71)	0.041	4.57 (3.02)	0.132	0.116
WOMAC-B (stiffness)	4.07 (3.20)	0.205	1.38 (3.32)	0.678	0.417
WOMAC-C (function)	5.44 (2.73)	0.047	2.16 (2.90)	0.457	0.116
Heel to buttocks (cm)	−0.30 (1.02)	0.770	−0.24 (1.03)	0.820	0.953
Knee circumference (cm)	0.02 (0.37)	0.967	−0.05 (0.44)	0.907	0.988
Maximum extension (degrees)	0.15 (0.56)	0.796	−0.49 (0.53)	0.356	0.386
Maximum flexion (degrees)	−0.34 (1.30)	0.795	−0.86 (1.37)	0.531	0.746

[a]SE = standard error.

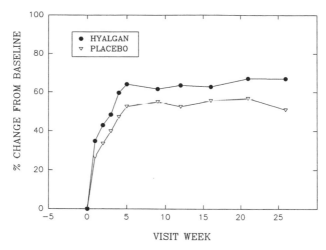

Fig. 1. VAS pain on 50-foot-walk test by week and treatment group for completed subjects: percentage change from baseline.

8.4. Primary Effectiveness Measures

1. VAS for pain, 50-foot-walk test: Among completers, the test group exhibited statistically significantly greater improvement in the VAS for pain, 50-foot-walk test as compared to the placebo group at wk 4, 5, 12, 21, and 26 ($p < 0.05$), and nearly significant differences at wk 3 ($p = 0.057$), wk 9 ($p = 0.114$), and wk 16 ($p = 0.111$). At wk 26, the difference between the test group and the placebo group-adjusted means was 8.85 mm ($p = 0.0043$), which is a difference of approximately one-third SD, and exceeds the first specified component of the treatment success criterion. In addition, improvement in pain on the VAS for the 50-foot-walk test exhibited by the test group exceeded that exhibited by the NSAID group at wk 26, although the difference was not statistically significant. This satisfies the third component of the success criterion.

    The time-course of the test product effect, illustrated in Fig. 1, indicates that pain relief is apparent after the first injection, reaches a plateau after the fifth injection, and persists unabated until the end of the observation period (sixth month).

    The test group's pain VAS score indicates that there was less pain in this group than in the NSAID group at wk 26, although the p-value for the difference was > 0.05. This exceeded the success criterion requirements.

2. Categorical assessment of pain—subjects: Pain was also assessed categorically by subjects using a scale of 0–5 (0, no pain and 5, disabled).

The subjects' assessment of pain indicated a shift toward less pain at wk 26 in all groups. A higher percentage of test subjects than placebo-treated subjects had "no pain" or "no pain or slight pain" at 26 wk. In addition, the percentage with "no pain" or "no pain or slight pain" in the test group is greater than the corresponding percentage in the NSAID group at wk 26. There was also a higher percentage of improvers in the test group as compared to both the placebo and the NSAID group. The results of the categorical assessment by subjects are concordant with those obtained in the VAS 50-foot-walk test pain assessment, and thus satisfy the second and third components of the treatment success criterion.

3. Categorical assessment of pain—masked observers: The masked observers' categorical assessment of pain generally agreed with the subjects' assessment of pain and indicated a shift toward less pain at wk 26 in all groups with higher percentage of test than placebo subjects having "no pain" or "no pain or slight pain." In addition, there was a higher percentage of improvers in the test group compared to both the placebo and NSAID group. The results of the categorical assessment by the masked observers are concordant with those obtained in the VAS 50-foot-walk test pain assessment, and thus satisfy the second and third components of the treatment success criterion.

8.5. Secondary Effectiveness Measures

1. The WOMAC scale: The WOMAC Osteoarthritis Index is a validated and reliable health status instrument used in the measurement of joint pain (WOMAC A), stiffness (WOMAC B), and function (WOMAC C) in subjects with osteoarthritis of the knee or hip in clinical trials (*see* Figs. 2–4).[4] Test subjects had the lowest mean VAS score among the three treatment groups at wk 26 for each major section of the WOMAC scale. The WOMAC A (pain) means for the test group were consistently lower than those of the NSAID group from wk 5–26. None of the differences in the WOMAC scale between the test group and the NSAID group had p-values < 0.05. However, the differences in effect size between test and placebo groups yielded p-values of 0.041 for WOMAC A and 0.047 for WOMAC C.

   Other secondary outcome measures (time to perform 50-foot-walk test, knee range of motion, heel-to-buttock distance, and effusion) had only modest changes, generally favorable to the test product. Overall, the observed differences in these parameters did not appear to be meaningful.

2. Acetaminophen consumption: Test-placebo group differences in acetaminophen consumption were not statistically significant; whereas both the placebo-group ($p = 0.055$) and the test-group consumptions ($p = 0.097$) were nearly significantly higher than the NSAID group consump-

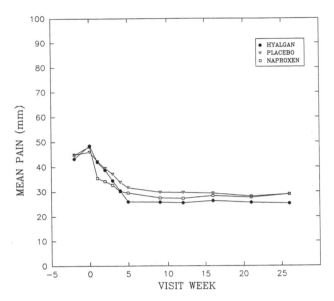

Fig. 2. Mean pain on WOMAC A scale by week and treatment group: completed subjects (mm).

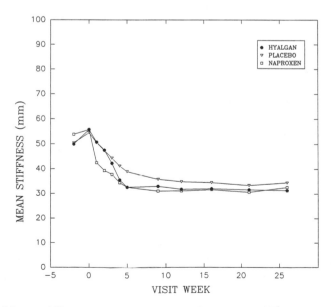

Fig. 3. Mean stiffness on WOMAC B scale by week and treatment group: completed subjects (mm).

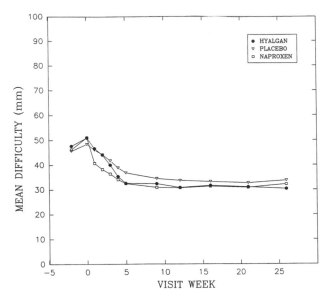

Fig. 4. Mean difficulty on WOMAC C scale by week and treatment group: completed subjects (mm).

tions (two-tailed T-test for equal variances). The NSAID subjects tended to consume approximately one tablet less per day than the other subjects.
3. Subject and masked-observer assessment of treatment effectiveness: Subject and masked-observer assessment of treatment effectiveness also satisfied the criterion for measuring success. A higher percentage of completers in the test group rated their treatment "very effective" than did the completers in the placebo-treated group. The same held true for the masked-observer assessments. In addition, the percentage of "not effective" ratings was lower in the test group than in the placebo group. The percentage of the placebo group for which the overall treatment effectiveness was rated as "very effective" by the subject or the masked-observed was 50% or more in completers. Despite this high rate, the test treatment elicited a higher percentage of favorable ratings than placebo.

8.6. Safety

1. Deaths: One subject (subject no. 08201), an 80-yr-old Caucasian woman from the test treatment group, died of a myocardial infarction in the hospital after 53 d on study. The death was considered not related to treatment and "probably" related to concomitant illness.
2. Discontinuations: As noted above, 162 subjects did not complete the study. The only statistically significant difference between treatment

Table 2
Distribution of Selected Adverse Events

Adverse event	Treatment			
	Test N (%)	Placebo N (%)	NSAID N (%)	Total N (%)
Local joint pain and swelling[a]	21 (12.8)	22 (13.1)	10 (6.1)	53 (10.7)
Ecchymosis (local)	11 (6.7)	10 (6.0)	16 (9.8)	37 (7.5)
Headache	30 (18.3)	29 (17.3)	17 (10.4)	76 (15.3)
Pain at injection site[b]	38 (23.2)	22 (13.1)	14 (8.6)	74 (14.9)
Pruritus (local)	12 (7.3)	7 (4.2)	7 (4.3)	26 (5.3)
Rash (local)	12 (7.3)	16 (4.2)	13 (8.0)	41 (8.3)
Gastrointestinal complaints[a]	42 (25.6)	50 (29.8)	60 (36.8)	152 (30.7)

[a] Statistically significant ($p < 0.10$); Fisher's exact test: Test vs placebo vs NSAID, subjects with events vs remainder of subjects.
[b] Statistically significant ($p < 0.001$); Fisher's exact test: Test vs placebo vs NSAID, subjects with events vs remainder of subjects.

groups with respect to reasons for discontinuation was the number of discontinuations resulting from gastrointestinal complaints in the NSAID group ($p < 0.0001$). The number of subjects who terminated because of injection-site pain was higher in the test treatment group compared to the other groups, but the difference was not statistically significant.

3. Adverse events: There were 2201 adverse event reports, of which 1114 were repeated reports of the same event in the same subject and 1087 were unduplicated reports. The frequency of adverse events reported across treatment groups was comparable. The adverse-event maximal severity distribution observed in the test and placebo groups was comparable, whereas fewer NSAID-treated subjects reported "severe" adverse events, although the difference was not statistically significant. Almost half of all subjects who reported that their adverse events were "probably related" to the investigational device were in the test group; this proportion was larger than placebo or NSAID groups. The proportion of adverse events that was "probably related" to the oral study medication was significantly higher in the NSAID group as compared to the other two treatment groups.

The distribution of the six most frequently reported adverse events, which comprise 75% of all adverse events, is summarized in Table 2. These events were local joint pain and swelling, ecchymosis, headache, pain at injection site, pruritus, rash, and gastrointestinal complaints. Test treatment subjects had the highest incidence of pain at injection site; the difference in incidence of this adverse event between the test group and either the NSAID or the placebo groups was statistically significant. There were no meaningful differences between groups with respect to the incidence of ecchymosis, pruritus, and rash.

Headache was least frequently reported among NSAID-treated subjects, and test and placebo groups had nearly equal percentages of subjects with headaches. The lower frequency in the NSAID-treated group may be a result of a systemic effect. Gastrointestinal adverse events were recorded more frequently in the NSAID-treated group as compared to the other treatments and the difference was statistically significant.

4. Severe events: A total of 81 subjects had adverse events classified "severe." The distribution of maximal severities indicates nearly identical distributions for test and placebo-treated subjects whereas fewer NSAID-treated subjects had events classified "severe."

5. Laboratory findings: A number of laboratory parameters exhibited statistically significant variations in each treatment group among the baseline, wk 9, and wk 24 evaluations. None of these changes was considered clinically significant.

9. Summary and Conclusions

In spite of the inherent assessment difficulties and the stringent success criterion, the design of this clinical trial proved to be capable of testing the hypothesis that a cycle of five injections of a hyaluronic acid product is clinically useful in the long-term treatment of osteoarthritis of the knee. The statistically significant difference in the VAS pain assessment, a measurement that quantifies a subjective outcome, between the test and the placebo-treated group was clinically significant. This conclusion was supported by the concordant results from additional measures used to assess effectiveness in this trial. Specifically, higher proportions of test subjects reported no or mild pain in the categorical assessment of pain as compared to the placebo-treated subjects, and more test subjects reported improvement in their pain than placebo-treated subjects. Furthermore, results reported from other controlled clinical trials in patients with osteoarthritis of the knee confirm that differences of similar magnitude in the VAS are viewed as clinically useful and significant. For example, Williams et al.[5] reported that a difference in mean values of 5.36 mm on the VAS scale for pain at rest between subjects treated with a total daily dose of 1000 mg naproxen and subjects treated with a total daily dose of 1300 mg acetaminophen for 6 wk was statistically significant and that the naproxen treatment was "superior to acetaminophen" for this outcome.

Additionally, secondary measures of effectiveness were concordant with the primary measures. The observed improvement in the WOMAC C, an assessment of function, is particularly noteworthy since it provides further evidence of the clinical utility of the test product in everyday activities of osteoarthritis patients.

The level of pain reduction observed in the saline-injected placebo group is probably partially attributable to the use of escape acetaminophen during the trial. Ethical considerations precluded the use of a true placebo for extended periods in a patient population with moderate to marked pain. Indeed, this arm of the trial was really an alternative treatment arm (i.e., acetaminophen), which is currently recommended as a first step therapy in this condition. A number of investigators have reported some degree of pain relief in patients with osteoarthritis who have been treated only with acetaminophen. For example, in a clinical trial where subjects with mild or moderate osteoarthritis of the knee were given 4000 mg acetaminophen/d for 4 wk, a 10% improvement in walking pain and a 6% improvement in resting pain (assessed by VAS score), a 23% improvement in the overall pain score (assessed by the Stanford Health Assessment Questionnaire disability and pain scale), and a global assessment of improvement by the physician in 37% of the acetaminophen-treated subjects (all as compared to baseline) were observed.[6] In addition, intra-articular injections of physiological saline have been shown to elicit a pain-relieving effect.[7, 8] Thus, the favorable results in the placebo group may be explained by a combination of acetaminophen dosing, the placebo effect of the intra-articular injection, aspiration of synovial fluid, and possibly the ability of saline to provide marginal lubrication. It is important to note that the test and the placebo-treated groups are comparable by design with respect to all these factors; both groups consumed the same amount of acetaminophen, were injected with physiological saline, and had synovial fluid aspirated. Because the mean consumption of acetaminophen by the test and placebo-treated subjects was comparable, the greater pain relief observed in the test subjects could not be attributed to differential use of the escape analgesic.

The comparison with NSAID was also instrumental in assessing the clinical value of the test product. The success requirement that this device had to be at least 50% as effective as an NSAID was based on the following considerations. NSAIDs are very effective and therefore commonly used in osteoarthritis, but their adverse reactions are frequent and potentially dangerous. An alternative therapy with diminished adverse reactions, particularly with respect to gastropathy, even only half as effective as an NSAID, would be of benefit to a large number of patients. In this context, the finding that the test product was superior to a widely used NSAID in many parameters is of particular importance. Thus, the choice of appropriate control groups in this study allowed the detection of a clinically useful effect of the test treatment while preserving the welfare of the study subjects.

In summary, design constraints of this trial included the evaluation of a subjective primary effectiveness parameter (pain), potential variability with respect to treatment and assessment, an ethical requirement to provide escape medication, and a strict success criterion. Appropriate choices of statistical methodology, assessment instruments, and control groups enabled these constraints to be overcome in this clinical trial and provide evidence of the clinical utility of a cycle of five intra-articular injections of a hyaluronic acid product.

References

1. Ohnhaus, E. E. and Adler, R. 1975. Methodological problems in the measurement of pain: a comparison between verbal rating scale and the visual analogue scale. *Pain* 1:379–384.
2. Sriwatanakul, K., Kelvie, W., Lasagna, L., Calimlim, J. F., Weis, O. F., and Mehta, G. 1983. Studies with different types of visual analog scales for measurement of pain. *Clin. Pharmacol. Ther.* 34:234–239.
3. Dougados, M., Nguyen, M., Listrat, V., and Amor, B. 1993. High molecular weight sodium hyaluronate (hyalectin) in osteoarthritis of the knee: a 1 year placebo-controlled trial. *Osteoarthritis Cart.* 1:92–103.
4. Bellamy, N. and Buchanan, W. W. 1993. Clinical evaluation in rheumatic diseases. In *Arthritis and Allied Conditions: A Textbook of Rheumatology* (McCarty D.J. and Koopman W.J. eds.). Lea & Febiger, Philadelphia, pp. 157–172.
5. Williams, H. J., Ward, J. R., Egger, M. J., Neuner, R., Brooks, R. H., Clegg, D. O., Field, E. H., Skosey, J. L., Alarcón, G. S., Willkens, R. F., Paulus, H. E., Russell, I. J., and Sharp, J. T. 1993. Comparison of naproxen and acetaminophen—a two-year study of treatment of osteoarthritis of the knee. *Arthritis Rheum.* 36:1196–1206.
6. Bradley, J. D., Brandt, K. D., Katz, B. P., Kalasinski, L. A., and Ryan, S. I. 1991. Comparison of an anti-inflammatory dose of ibupro-

fen, an analgesic dose of ibuprofen, and acetaminophen in the treatment of patients with osteoarthritis of the knee. *N. Engl. J. Med.* 325: 87–91.

7. Formiguera Sala, S. and Esteve de Miguel, R. 1995. Intra-articular hyaluronic acid in the treatment of osteoarthritis of the knee: a short term study. *Eur. J. Rheumatol. Inflamm.* 15:33–38.
8. Huskisson, E. C. 1995. Randomized, placebo-controlled study to compare the effectiveness of, and the patient satisfaction with intra-articular HYALGAN® in the treatment of osteoarthritis of the knee. Final report dated 12 June.

11

Role of Device Retrieval and Analysis in the Evaluation of Substitute Heart Valves

Frederick J. Schoen

1. Introduction

Surgical replacement of diseased valves with functional substitutes is the dominant therapeutic modality in patients with symptomatic valvular heart disease, and improves the survival and enhances the quality of life of many individuals.[1] Nevertheless, problems associated with prostheses remain a major impediment to the long-term success of this procedure. Despite considerable improvement in the technology of heart valve replacement devices since their first successful use approx 35 yr ago, both mechanical and tissue heart valve substitutes (*see* Fig. 1) remain imperfect, and prosthesis-associated complications have considerable impact on the long-term outlook for persons who have had valve replacement surgery.

Reoperation or postmortem examination of patients who have had valve replacement, and preclinical animal valve implant studies provide specimens of prostheses whose analysis can lead to valuable data and insights. Pathological evaluation of substitute heart valves has:

1. Contributed to the care of valve replacement patients;
2. Established the rates, morphology, and mechanisms of prosthesis-associated complications;
3. Elucidated the structural basis of favorable valve performance;

From: *Clinical Evaluation of Medical Devices: Principles and Case Studies*
Edited by K. B. Witkin Humana Press Inc., Totowa, NJ

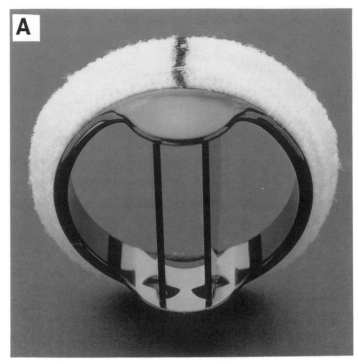

Fig. 1. Mechanical prosthetic and tissue bioprosthetic heart valve replacement devices. (**A**) St. Jude Medical carbon bileaflet tilting disk prosthesis, the most widely used mechanical heart valve prosthesis. Courtesy St. Jude Medical Inc. (St. Paul, MN).

4. Predicted the effects of developmental modifications on safety and efficacy; and
5. Enhanced our understanding of patient–prosthesis and blood–tissue-biomaterial interactions.

In this chapter, emphasis will be on the rationale and overall contributions of hypothesis-driven explant analysis of valve substitutes applicable to failed as well as nonfailed and, to a large extent, unimplanted specimens. These aspects have not been previously described in detail. In contrast, technical approaches and procedures useful in both preclinical and clinical implant retrieval have been widely documented.[2-7]

2. General Considerations

The goals of routine hospital surgical pathology or autopsy examination of an artificial valve are generally restricted to documentation

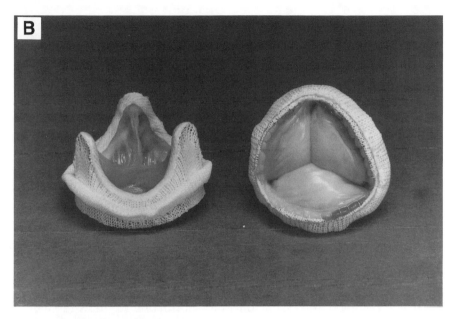

Fig. 1. (cont'd). **(B)** Carpentier-Edwards porcine aortic bioprosthetic valve, the most widely used type of tissue heart valve prosthesis. Courtesy Edwards Division, Baxter HealthCare Corp. (Santa Ana, CA).

of the specific valve type that has either been removed at reoperation or the patient has died with, and diagnosis of a clinical abnormality that requires therapeutic intervention, such as a valve-related infection (prosthetic valve endocarditis). Detailed correlation of morphologic features with clinical signs, symptoms, and dysfunctional physiology is usually not performed. However, directed and informed pathological examination of cardiac valve prostheses retrieved during preclinical animal studies or at reoperation or autopsy of human patients can provide valuable additional information. First, preclinical studies of modified designs and materials are crucial to developmental advances. These investigations usually include implantation of functional devices in the intended location in an appropriate animal model, followed by noninvasive and invasive monitoring, followed by specimen explantation and detailed pathological analysis. Second, for individual patients, demonstration of a propensity toward accelerated calcification or a predisposition to hypercoagulability as a cause of thrombosis would impact greatly on further management. Third, clinicopathologic analysis of cohorts of patients who have received a new or modified valve prosthesis type evaluates its safety and efficacy to an extent beyond that obtainable by either in vitro tests of durability and biocompatibility or preclinical investigations of valve implant configurations in large animals. Moreover, through analysis of rates and modes of failure as well as morphologic and mechanistic characterization of specific failure modes in patients with implanted medical devices, retrieval studies can contribute to the development of methods for enhanced clinical recognition and elucidation of the pathogenesis of failure mechanisms that guide future development of improved prosthetic devices to eliminate complications. Emphasis is usually directed toward failed valves; however, careful and sophisticated analysis of removed prostheses that are functioning properly, and indeed, detailed analyses of preimplantation structural features and their evolution following implantation, can yield an understanding of structural correlates of favorable performance and identify predisposition to specific failure modes, such as thrombosis or mechanical failure.

Device retrieval analysis also has an important regulatory role, as specified in the Safe Medical Devices Act of 1990 (PL101-629), the first major amendment to the Federal Food, Drug and Cosmetics Act since the Medical Device Amendments of 1976.[8, 9] The user-requirements of the recent legislation require health-care personnel and hospitals to report (within 10 d) to the Food and Drug Administration (FDA) or manufacturers or both (depending on the nature of the occurrence) all prosthesis-associated complications that cause death, serious illness, or injury. Such incidents are often discovered during a pathologist's diagnostic evaluation of an implant in the autopsy suite or the surgical pathology laboratory.

The ongoing emphasis on health-care financing concerns may cause a broadening of the concept of patient–prosthesis matching to include the relative rates and varying nature of complications of different devices of disparate cost. Analysis of patients and prostheses provides important data that can be used to approach the (justifiably controversial) question: "Can less expensive devices with adequate performance provide sufficient benefit in particular patient populations, to obviate the use of high-performance, but higher cost implants, where they may be unnecessary?" For example, can an inexpensive heart valve prosthesis with an expected 20-yr lifetime be used in an octogenarian, thereby reserving more costly devices with an estimated 50-yr lifetime for younger, more active patients?

Table 1
Objectives of Substitute Heart Valve Retrieval and Analysis

Establish rates, modes, and mechanisms of failure
Enhance patient management by surveillance for and noninvasive
recognition of complications
Enhance identification of patient influences on device function
Enhance device selection and patient–prosthesis matching criteria
Establish structural correlates of favorable performance
Eliminate complications by device development
Predict effects of prosthesis modifications on efficacy and safety
Identify subclinical patient–prosthesis interactions
Elucidate blood–tissue–biomaterials interaction mechanisms

Table 2
Features of Successful Implant Retrieval Studies

Activity is hypothesis-driven
Consideration of known and potential failure modes of specific devices
and settings
Knowledge of pertinent clinical data
Broad inputs from all concerned disciplines, including a pathologist
Stratified analyses (mandatory vs elective), with availability of expert
laboratories for specialized/advanced analyses
Recognition that some analyses may be mutually exclusive, with material
taken and prepared for all reasonable analysis possibilities
Data recorded on carefully designed, study-specific forms
Data quantification, with appropriate statistical analyses, wherever possible

The major objectives of substitute heart valve retrieval and analysis
are summarized in Table 1. Examination of substitute heart valves
has several necessary sources of input as well as conditions; these may
vary according to the specific goals of the evaluation. The essential
and desirable components of and prerequisites for high quality implant retrieval analysis are summarized in Table 2. Since failure modes
of prosthetic heart valves depend on the device model and type, patient
factors, and the site of implantation, both experimental and clinical
analyses require knowledge of established and potential failure modes
of various devices in particular situations as well as the clinical data
pertinent to specific cases. Extensive information is available on the
failure modes and pathological features associated with the many different generic types and models of heart valve substitutes that have

been used experimentally and clinically.[10-15] Large scale experimental or clinicopathologic investigations should also benefit from an analytical protocol formulated with input from all relevant specialists, including a pathologist, carefully designed, study-specific data recording forms, and wherever possible, quantitative data measurement and use of appropriate statistical analyses. Moreover, evaluation techniques should be stratified. Level I studies include routine evaluation modalities capable of being done in virtually any laboratory, and that characterize the overall safety and efficacy of a device, including complications, cause of death, and critical blood–tissue–biomaterial and patient–device interactions (such as gross and dissecting microscope examination, photography, microbiologic cultures, histology, and, in some cases, radiography of the specimen). Level II studies comprise well-defined and meaningful test methods that are either difficult or expensive to perform, require special expertise, or yield more investigative or esoteric data, such as scanning or transmission electron microscopy, or chemical, biochemical, immunological, or molecular techniques (e.g., calcium assay, protein measurement, immunoperoxidase localization in tissue sections of a protein for which an antibody is available; or *in situ* hybridization to localize messenger RNA in tissue, as an indication of specific cellular gene expression). Since some Level II analyses may be mutually exclusive, some material might routinely need to be accessioned, set aside (during Level I analyses), and prepared for more specialized Level II studies, in the event that they should be indicated later. Prioritized, practical approaches using these guidelines have been described for heart valves, cardiac assist devices and artificial hearts, and other cardiovascular devices.[2-7]

3. Preclinical Implant Retrieval

Procedures used to evaluate cardiovascular devices and prostheses after function in animals and humans are largely the same. However, subject to humane treatment considerations, enumerated in institutional and National Institutes of Health (NIH) guidelines that enforce the Federal Animal Welfare Act of 1992, animal studies permit more detailed monitoring of device function and enhanced observation of morphologic detail (including blood-tissue-biomaterials interaction), as well as frequent assay of laboratory parameters (such as indices of platelet function or coagulation), and allow *in situ* observation of fresh implants following elective sacrifice at desired intervals. In addition, specimens from experimental animals are often obtained rapidly, thereby minimizing the autolytic changes that occur when tissues are removed from their blood supply. Furthermore, advantageous technical adjuncts may be available in animal but not human investigations, including in vivo studies, such as injection of radiolabeled ligands for imaging platelet deposition,[16] fixation by pressure perfusion that maintains tissues and cells in their physiological configuration following removal,[17, 18] and injection of various substances that serve as informative markers during analysis (such as indicators of endothelial barrier integrity).[7, 19] Animal studies often facilitate observation of specific complications in an accelerated time frame, such as calcification of bioprosthetic valves, in which the equivalent of 5–10 yr in humans is simulated in 4–6 mo.[20, 21] Moreover, in preclinical studies, experimental conditions can be held constant among groups of subjects with the same valves, including nutrition, activity levels, and treatment conditions. Consequently, concurrent control implants, in which only a single critical parameter varies, are often available in animal but not human studies.

4. Clinical Implant Retrieval

Clinical and pathological studies have demonstrated that virtually all types of widely used cardiac valve substitutes suffer deficiencies and complications that have limited their success.[10–15] Indeed, prosthesis-associated pathology is a major determinant of the prognosis of patients who have had valve replacement. Among patients who die following valve replacement, the immediate cause of death is device-related in 25–61%. Moreover, clinical investigations of individual valve types in randomized studies show that more than 60% of valve replacement patients suffer an important adverse prosthesis-associated event within 10 yr of surgery, irrespective of valve type.[12, 13] However, patient outcome after cardiac valve replacement also depends on both irreversible cardiac pathology secondary to the original valve disease (especially left ventricular myocardial hypertrophic and degenerative changes) and other cardiac pathology that occurs even in nonvalve replacement patients, such as coronary arterial atherosclerotic occlusions.[3, 15, 18] This reality underscores the importance of examining a valve or other device in the functional anatomic context (e.g., the heart and patient), and the need to perform careful and complete animal necropsy or human autopsy of subjects, if at all possible.

Table 3
Complications of Cardiac Valve Substitutes[a]

Generic	Specific
Thrombotic limitations	Thrombosis
	Thromboembolism
	Anticoagulation-related hemorrhage
Infection	Prosthetic valve endocarditis
Structural dysfunction	Wear
	Fracture
	Poppet escape
	Leaflet immobility
	Cuspal tear
	Calcification
	Commissural region dehiscence
Non-structural dysfunction	Pannus (tissue overgrowth)
	Entrapment by suture or tissue
	Paravalvular leak
	Disproportion
	Hemolytic anemia
	Noise

[a]Modified by permission from ref. *14*.

Valve-related complications are categorized as thromboembolism and related problems, infection, structural dysfunction, and nonstructural dysfunction, as summarized in Table 3. Although overall rates of valve-related complications are similar for mechanical prostheses and bioprostheses, the frequency and nature of specific valve-related complications vary with the prosthesis type, model, site of implantation, and patient characteristics. Contemporary mechanical prostheses are durable with a few notable exceptions; however, such valves are prone to thrombosis and thromboembolism (*see* Fig. 2), necessitating chronic anticoagulation in patients who have received them. In contrast, tissue valves have a relatively low rate of thromboembolism without anticoagulant therapy, but virtually all bioprostheses used to date have had limited durability, nearly exclusively because of cuspal degeneration (primary tissue failure with calcification and tearing), exemplified by glutaraldehyde-pretreated porcine aortic valves (*see* Fig. 3).[3, 11, 15] In several valve types with consistent failure modes, detailed pathologic analysis coupled with clinical data have implicated specific causal mechanisms of deleterious interaction.

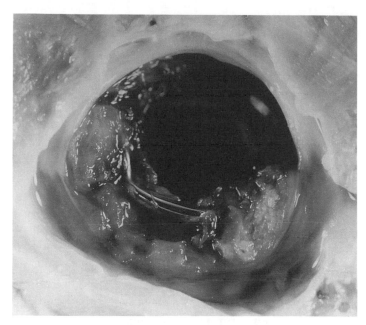

Fig. 2. Thrombotic occlusion of a mechanical heart valve prosthesis viewed from distal (outflow) aspect. Reproduced by permission from ref. *14.*

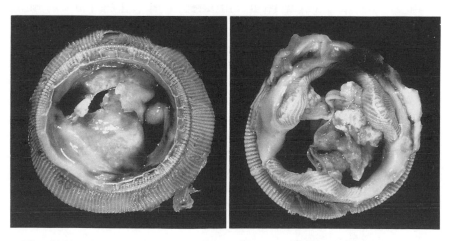

Fig. 3. Porcine bioprosthetic valve, failing by calcification with secondary cuspal tears. **(Left)** Inflow aspect. **(Right)** Outflow aspect.

5. "Case" Histories
Demonstrating the Utility of Retrieval Analysis

The literature contains numerous instances where problem-oriented implant analysis studies have yielded important insights. Several are summarized below.

5.1. Braunwald-Cutter Cloth-Covered Heart Valve

Introduced in the early 1970s, the Braunwald-Cutter cloth-covered caged-ball mechanical prosthetic heart valve had a ball fabricated from silicone and an open-cage apex that was covered by polypropylene mesh.[22] The major innovation was the cloth-covered cage struts, intended to encourage tissue ingrowth and thereby decrease thromboembolism.

Preclinical studies of the Braunwald-Cutter valve concept utilized mitral implants in pigs, sheep, and calves; in such models the cloth-covered struts were rapidly and appropriately healed by endothelium-coated fibrous tissue (a neointima).[23, 24] However, subsequent clinical studies of these valves demonstrated that, similar to other cloth-covered heart valves, cloth wear was abundant in both mitral and aortic prostheses (*see* Fig. 4).[25] Some patients had sufficient cloth wear accompanied by abrasive wear of the poppet that the ball escaped through the spaces between cage struts, a complication that would be rapidly fatal without emergency surgery. In the aortic position, both cloth and ball changes were accentuated, and ball escape was more frequent.

Detailed and formal analysis of clinically removed mitral and aortic Braunwald-Cutter prostheses retrieved at various institutions included semiquantitative and quantitative characterization of strut coverage and poppet wear to elucidate the mechanisms of the disparate clinico-pathologic behavior between mitral and aortic sites. Tissue ingrowth was usually present on the struts of the mitral prosthesis (at least in part) after prolonged periods of implantation, but the fabric on aortic valves was not covered by tissue rapidly enough to prevent excessive poppet/cloth abrasive wear.[26] These data also suggested that mitral Braunwald-Cutter prostheses need not be electively replaced without specific indication. This case example demonstrates that

1. Human trials and extensive use may reveal important complications not predicted by animal investigations;

Fig. 4. Braunwald-Cutter mitral heart valve prosthesis, demonstrating marked wear of the cloth covering the struts.

2. The more vigorous healing that occurred in animals rather than humans prevented preclinical prediction of the problem that occurred in people;
3. The consequences of valve failure can depend exclusively on location of implant;
4. Data quantitation may facilitate the understanding of a failure mode; and
5. The results of implant retrieval studies can impact on patient management.

5.2. Björk-Shiley 60–70° Convexo-Concave (C-C) Heart Valve

An unusually large cluster of a frequently fatal complication of the widely used Björk-Shiley C-C heart valves with opening angles of 60 or 70° has been reported in which the welds anchoring the metallic outlet strut to the housing have fractured, leading to separation from the housing and consequent disk escape with acute valve failure (*see* Fig. 5). As of August 1994, this complication had occurred 557 times among approx 81,000 valves of this type implanted world-wide during the period 1979–1986; approx 46,000 individuals currently live with this valve.[27] Clinical studies have identified large valve size, mitral site

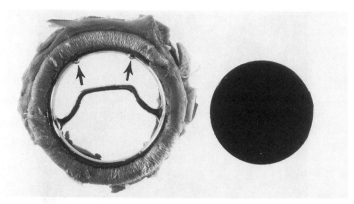

Fig. 5. Björk-Shiley heart valve prosthesis with fracture of the lesser (outflow) struts previously welded to the metal frame. Fracture surfaces are noted by arrows. The fractured strut could not be located at autopsy. Reproduced by permission from ref. *14*.

of implantation, recipient age > 50 yr old, and valve manufacture in 1981 and early 1982 as risk factors for this mode of dysfunction.[28-30]

Pathology studies revealed that the underlying problem is metal fatigue resulting from overrotation of the disk during valve closure.[14] The design can cause an abnormally hard strike of the disk on the metallic outlet strut, leading to excessive bending stresses at or near the welds joining the outlet struts to the housing, potentially coupled with intrinsic flaws of the welds. In support of this failure mechanism, retrieved Björk-Shiley valves of this design often demonstrate a pronounced wear facet at the tip of the outlet struts. In some cases, the excessive contact force between the disk and the tip of the outlet strut is manifest by localized pyrolytic carbon wear deposits at sites of apposition. Scanning electron microscopy of the fractured surfaces demonstrated that fractures begin on the inlet side of the outlet strut and suggested that the first strut leg fracture typically initiates at or near the point of maximum bending stress in the center of the inlet strut leg-annulus junction. The crack origin site in the second strut leg to fracture is often rotated slightly toward the first, since the remaining intact strut leg is subjected to both bending and contortion after the first fracture occurs. The initiation site could be traced to a site of weld shrinkage porosity and/or inclusion in most cases. Occasionally, valves with only a single strut fracture are encountered.[31]

In animal studies in which Björk-Shiley 60–70° C-C valves were implanted in sheep and instrumented with strain gages, it was shown that impact forces vary greatly with the cardiac activity and that such loads occurring during exercise significantly exceeded those measured under sedentary conditions. An increase in the speed and intensity of left ventricular contraction (measured as dP/dt) and elevated left ventricular pressure at valve closing augmented impact forces. Thus, rise of the left ventricular pressure during early systole seems to govern the closing dynamics of Björk-Shiley C-C mitral valves and the impact forces are in good agreement with those predicted by theoretical models. Moreover, the concept that hyperdynamic cardiac activity is contributory to catastrophic failure is supported by a recent clinical analysis that showed higher risk of fracture with increased cardiac output.

This situation demonstrates that elucidation of a failure mode by detailed materials failure analysis and effective use of carefully designed animal experiments can have potential impact on patient management. Understanding this mode of failure justifies development of noninvasive testing modalities (via high definition radiography[32] or acoustic characterization of strut status[33]) to establish when one strut has fractured prior to the onset of clinical failure and to caution patients with mitral valves against activities that enhance dP/dt. Such patients might also be treated with drugs (such as β blockers) that reduce dP/dt.

5.3. Bioprosthetic Heart Valve Calcification

Calcification is an important pathologic process contributing to failure of bioprostheses fabricated from porcine aortic valves (*see* Fig. 3).[3, 11, 14] Calcific failure occurs as early as 4 yr postoperatively in adults, averages approx 7–8 yr, and occurs more frequently and earlier yet in children and adolescents. Studies of retrieved experimental and clinical implants have characterized calcification-induced failure modes,[34] patterns and extent of mineral deposition,[35] the nature of the mineral phase,[36] and early calcification events.[37, 38]

Further study directed to mechanisms and prevention have utilized calcification of bioprosthetic tissue in both circulatory and subcutaneous (sc) experimental models, both of which have morphologic features similar to those observed in clinical specimens. However, experimental progression of calcification is markedly accelerated. Valves implanted as mitral replacements in sheep calcify extensively in 3–4 mo and sc implants of bioprosthetic tissue in rats achieve calcium levels comparable to those of failed clinical explants in 8 wk or less.

We have utilized the sc implantation model extensively as a technically convenient, economically advantageous, and quantifiable model for investigating host and implant determinants and pathobiology of mineralization, as well as for screening and understanding the mechanisms of potential strategies for mineralization inhibition. Clinical and experimental studies indicate that calcification of bioprosthetic valves depends on host, implant, and mechanical factors.[38, 40]For example,

1. Pretreatment of tissue with an aldehyde crosslinking agent (such as glutaraldehyde) potentiates mineralization;
2. Calcification is most pronounced in areas of leaflet flexion, where deformations are maximal; and
3. Calcification is accelerated in children or young experimental animals.

Thus, the fundamental mechanisms for bioprosthetic tissue mineralization depend on specific biochemical modification of implant microstructural components induced by aldehyde pretreatments. The earliest mineral deposits in both clinical and experimental bioprosthetic tissue are localized to transplanted connective tissue cells; collagen involvement occurs later, possibly by an independent mechanism. Mineralization of connective tissue cells of bioprosthetic tissue appears to result from glutaraldehyde-induced cellular devitalization and the resulting disruption of cellular calcium regulation.[37, 38]

These pathobiology studies have provided the means to test approaches to reduce bioprosthetic valve failure by modifying host, implant, or mechanical influences. Most of the strategies for preventing bioprosthetic tissue mineralization involve modifications of either valve design or preparation details, or the local environment of the implant. Mechanisms of calcification inhibition by antimineralization treatments that have been considered and/or clinically investigated and/or experimentally investigated include (but are not limited to) extraction of calcifiable material, ionic and/or macromolecular binding to nucleation sites, and interference with calcium phosphate crystal growth.[41] Although durability of modified bioprosthetic heart valves can be best assessed using long-term clinical evaluation, an appropriate experimental testing program for antimineralization strategies can demonstrate essential features of efficacy and safety. Studies suggest that preclinical determination of the efficacy and safety of antimineralization treatments require at least four components:

1. Qualification (and wherever possible, assessment of mechanism) using sc implantation in rats;
2. Hydrodynamic/durability testing to show lack of excessive obstruction or regurgitation, as well as verify that no new failure modes are evident;
3. Morphologic studies of unimplanted valve material to assess the potential for treatment-induced degradative changes[42]; and
4. Valve replacement in the appropriate configuration and site in an animal model, usually juvenile sheep.

Thus, in the context of the present discussion, retrieval studies of bioprosthetic heart valves emphasize that biological failure mechanisms can be understood using specifically designed animal models, guided by and correlated with the results of studies of retrieved clinical specimens; and implant retrieval studies can be used as a critical component of a thoughtful testing program to assure efficacy and safety of potential therapeutic modifications.

5.4. Beall Disk Heart Valve Prosthesis and the Durability of Pyrolytic Carbon

The Beall heart valve prosthesis, introduced in 1967, was originally a low-profile disk valve composed of a disk fabricated from extruded Teflon and metal struts coated with Teflon; this design and material were intended to maximize thromboresistance. Following realization of poor wear properties of the disk, with sequelae of disk abrasion, including hemolysis and abnormal disk motion[43] (*see* Fig. 6), its composition was changed to a denser compression-molded Teflon. Nevertheless, this prosthetic design continued to exhibit wear-related problems,[44] and a new model of the valve with disk and struts fabricated from pyrolytic carbon was introduced in the early 1970s.

Presently, as a result of favorable mechanical and biological properties of pyrolytic carbon, nearly all mechanical heart valve prostheses in use have pyrolytic carbon occluders, and some have both carbon occluders and carbon cage components. These carbons, produced by pyrolyzing a gaseous hydrocarbon in a fluidized bed chamber, exist in a poorly crystalline form with properties regulated by specific processing conditions. Laboratory studies have shown that the fracture stress of pyrolytic carbon is very high, as used in valves the material does not undergo critical degradation with repeated cycling (fatigue), and these carbons have an exceptional resistance to abrasive wear.

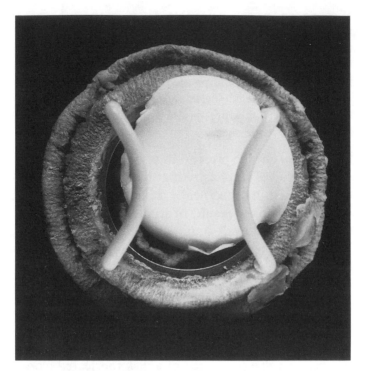

Fig. 6. Severe abrasive wear of disk poppet of Beall Teflon caged-disk mitral valve prosthesis.

Additional advantages of pyrolytic carbon as a material for construction of heart valve components are its thromboresistance and ability to be fabricated into a wide variety of shapes.

In the early 1980s, clinical experience suggested that pyrolytic carbon contributed to a major advancement in the durability of prosthetic heart valves; however, this had not been verified by direct valve observation. To confirm the anticipated favorable wear resistance of pyrolytic carbon in the clinical environment, we recovered at necropsy or surgery and analyzed by surface scanning electron microscopy and surface profilometry eight carbon-containing mechanical valve prostheses, including several carbon Beall valves, implanted for 32 to 48 mo.[45] None of the prostheses had clinical or gross pathological malfunction or abrasive wear, but the linear contacts had some evidence of strut wear by scanning electron microscopy. No appreciable wear on carbon valve occluders was demonstrated by analytical surface profilometry. Our study suggests that the use of pyrolytic carbon as an

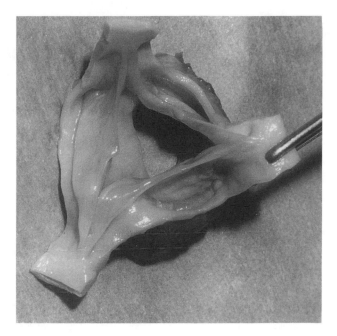

Fig. 7. Cryopreserved/thawed allograft heart valve. Removed from one individual who died, this aortic valve is ready for implantation in another as an aortic valve replacement. Courtesy CryoLife, Inc. (Marietta, GA).

occluder and as a strut material for mechanical heart valve prostheses has minimized progressive abrasive wear as a long-term complication of cardiac valvular replacement. The favorable clinical durability of pyrolytic carbon has subsequently been well documented.[46]

Analysis of implanted medical devices has traditionally concentrated on those devices that failed in service and paid insufficient attention to those serving the patient until death or removal from unrelated causes. The study described above emphasizes that detailed examination of functional (not failed) prostheses recovered from patients after long duration of implantation may yield worthwhile data, provided that a focused question is asked of the material.

5.5. Cryopreserved Allograft Heart Valves

Allografts are valves removed at either autopsy or heart transplantation from one human individual and transplanted to another. Today, most allografts are cryopreserved; i.e., stored at − 170°C in dimethyl sulfoxide (DMSO) until needed (*see* Fig. 7). With exceptionally good hemodynamic profiles low thromboembolic rates despite the absence of coagulation and low rate of infection, cryopreserved excised human heart valves have been increasingly used as valve replacements.[47, 48] Presently used allografts appear to have enhanced durability with high rates of freedom from degeneration,[49] but neither the structural basis of performance nor the pathophysiology of failure have previously been studied. There has been ongoing and vigorous debate regarding modes of failure, cellular viability, durability of the extracellular matrix, and the contribution to failure of immune responses.

We recently examined 20 explanted cryopreserved valves in place for several hours to 9 yr. These valves were obtained by virtue of the author's role as the core pathologist for a large clinical trial conducted under an Investigational Device Exemption (IDE) held by a consortium of five tissue banks. The original surgeries and explants were done among over 100 hospitals. Arrangements were made by the consortium to have removed valves sent directly to the core pathology laboratory, accompanied by clinical data. These were compared with both thawed but unimplanted allografts (obtained via the same route), and 16 donor aortic valves obtained from heart transplant recipients who later died at the Brigham and Women's Hospital.

Our studies indicate that removed cryopreserved allograft heart valves demonstrate progressively severe loss of the normal layered structure and stainable deep connective tissue cells with minimal inflammation.[50] Following either short-term or extended function, cryopreserved allograft heart valves have minimal, if any, viable cells, but largely retain the original collagen network; it is this preservation of the autolysis-resistant collagenous skeleton that likely provides the structural basis for function. Since inflammation in these valves is minimal in almost all cases, immune responsiveness likely has little impact on late allograft function or degeneration. Consequently, these valves, proven nonviable by morphological means and whose functionality seems primarily related to the largely preserved collagen, are unlikely to have the capacity to grow, remodel, or exhibit active metabolic functions. Interestingly, and in contrast, aortic valves of transplanted whole hearts maintain near-normal overall architecture, endothelial cells, and deep connective tissue cells.

Our studies of retrieved allograft valves emphasize that not only failure modes but also mechanisms of successful function can be elucidated by careful study, correlation of results from multiple types of implants can be exceedingly important in understanding structure–

Table 4
Problems in Clinical Device Retrieval

Lack of implant registry (data on patients, prosthesis types, complications)
Pathologist doing autopsy may not be aware that implant is in place
Retrieval requires autopsy/surgical removal from an informative anatomic
context
Removal-induced artifacts
Overinterpretation of pathologic findings
Disposition of specimen (patient, manufacturer, FDA, attorney, pathology
laboratory, research laboratory)
Inadequate clinical data
Potential biohazard/cleaning/packaging for transport/mode of discard
Priority of destructive vs nondestructive analyses
Inadequate funding for implant retrieval activities
Legal/ethical concerns (informed patient consent, confusion concerning
implant ownership, confidentiality of information)

function correlations, and a well-organized network that pools relatively unusual specimens obtained from many centers at an interested and competent analysis facility can lead to insights not readily approachable by clinical material that would accrue at a single institution.

6. Conclusions and Unresolved Issues

Pathological analysis of removed valve substitutes contributes to patient management and device development. Implant retrieval and analysis studies serve an important role in the evaluation of new and modified medical devices and may contribute to the management of individual patients. Optimal evaluation is driven by hypotheses and specific questions, utilizes technical steps appropriate for specific objectives, considers relevant failure modes, maximizes pertinent clinical data, and minimizes pitfalls. Nevertheless, difficulties remain, as summarized in Table 4, including most importantly:

1. Difficulty in tracking implants;
2. Pathological analysis requiring autopsy or surgical removal;
3. Legal/ethical issues, including ownership and confidentiality;
4. Technical problems (including removal-induced artifacts and overinterpretation of findings);
5. Frequently inadequate clinical data; and
6. Disinterest of conventional funding sources.

It is hoped that enhanced appreciation of the value of this activity will serve to facilitate the development of cooperation and collaborations that maximize the quantity and quality of the most useful specimens and relevant clinical data, as well as stimulate funding mechanisms.

7. Disclosure Statement

In the course of consultations on developmental and/or clinical research over approximately the past 5 years, Dr. Schoen has received or may receive something of value from the following organizations whose work is germane to the subject of this presentation: Advanced Tissue Sciences, Inc.; Allograft Heart Valve Tissue Bank Consortium; Autogenics, Inc.; Baxter, Inc.; BioMedical Design, Inc.; Bravo Cardiovascular, Inc.; CarboMedics, Inc.; CryoLife Cardiovascular, Inc.; Meadox Medical, Inc.; Medtronic, Inc.; Mitroflow Medical, Inc.; St. Jude Medical, Inc.; Symbion, Inc.; ThermoCardiosystems, Inc.

References

1. Rahimtoola, S. H. 1989. Perspective on valvular heart disease: an update. *J. Am. Coll. Cardiol.* 14:1–23.
2. Schoen, F.J. 1986. Special considerations for pathological evaluation of explanted cardiovascular prostheses. In *Handbook of Biomaterials Evaluation* (von Recum A. F. ed.). Macmillan, New York, pp. 412–435.
3. Schoen, F. J. 1989. *Interventional and Surgical Cardiovascular Pathology: Clinical Correlations and Basic Principles.* WB Saunders, Philadelphia, pp. 1–415.
4. Schoen, F. J., Anderson, J. M., Didisheim, P., Dobbins, J. J., Gristina, A. G., Harasaki, H., and Simmons, R. L. 1990. Ventricular Assist Device (VAD) pathology analyses: guidelines for clinical studies. *J. Appl. Biomat.* 1:49–56.
5. Anderson, J. M. 1993. Cardiovascular device retrieval and evaluation. *Cardiovasc. Pathol.* 2:199S–208S.
6. Schoen, F. J. 1995. Approach to the analysis of cardiac valve prostheses as surgical pathology or autopsy specimens. *Cardiovasc. Pathol.* 4:241–255.
7. Schoen, F. J. 1996. Pathological evaluation of explanted cardiovascular prostheses. In *Handbook of Biomaterials Evaluation* (von Recum A. F. ed.). 2nd Ed. Taylor & Francis Washington, in press.
8. Savage, R. A. 1991. New law to require medical device injury report. *CAP Today,* July, p. 40.
9. Kahan, J. S. 1991. The Safe Medical Devices Act of 1990. *Med. Dev. Diag. Ind.* Jan, 67.

10. Akins, C. W. 1995. Results with mechanical cardiac valvular prostheses. *Ann. Thorac. Surg.* 60:1836–1844.

11. Turina, J., Hess, O. M., Turina, M., and Krayenbuehl, H. P. 1993. Cardiac bioprostheses in the 1990s. *Circulation* 88:775–781.

12. Bloomfield, P., Wheatley, D. J., Prescott, R. J., and Miller, H. C. 1991. Twelve-year comparison of a Björk-Shiley mechanical heart valve with porcine bioprostheses. *N. Engl. J. Med.* 324:573–579.

13. Hammermeister, K. E., Sethi, G. K., Henderson, W. G., Oprian, C., Kim, T., Rahimtoola, S. 1993. A comparison of outcomes in men 11 years after heart-valve replacement mechanical valve or bioprosthesis. *N. Engl. J. Med.* 328:1289–1296.

14. Schoen, F. J., Levy, R. J., and Piehler, H. R. 1992. Pathological considerations in replacement cardiac valves. *Cardiovasc. Pathol.* 1:29–52.

15. Schoen, F. J. 1995. Pathological considerations in replacement heart valves and other cardiovascular prosthetic devices. In *Cardiovascular Pathology: Clinicopathologic Correlations and Pathogenetic Mechanisms* (Schoen F. J. and Gimbrone M. A. eds.). Williams and Wilkins, Baltimore, pp. 194–222.

16. Palatianos, G. M., Dewanjee, M. K., Panoutsopoulos, G., Kapadvanjwala , M., Novak, S., and Sfakianakis, G. N. 1994. Comparative thrombogenicity of pacemaker leads. *PACE Pacing Clin. Electrophysiol.* 17:141–145.

17. Davis, P. F. and Bowyer, D. E. 1975. Scanning electron microscopy: arterial endothelial integrity after fixation at physiological pressure. *Atherosclerosis* 21:463–469.

18. Clowes, A. W., Gown, A. M., Hanson, S. R., and Reidy, M. A. 1985. Mechanisms of arterial graft failure. 1. Role of cellular proliferation in early healing of PTFE prostheses. *Am. J. Pathol.* 118:43–54.

19. Reidy, M. A., Chao, S. S., Kirkman, T. R., and Clowes, A. W. 1986. Endothelial regeneration. VI. Chronic nondenuding injury in baboon vascular grafts. *Am. J. Pathol.* 123:432–439.

20. Levy, R. J., Schoen, F. J., Levy, J. T., Nelson, A. C., Howard, S. L., and Oshry, L. J. 1993. Biologic determinants of dystrophic calcification and osteocalcin deposition in glutaraldehyde-preserved porcine aortic valve leaflets implanted subcutaneously in rats. *Am. J. Pathol.* 113:143–155.

21. Schoen, F. J., Hirsch, D., Bianco, R. W., and Levy, R. J. 1994. Onset and progression of calcification in porcine aortic bioprosthetic valves implanted as orthotopic mitral valve replacements in juvenile sheep. *J. Thorac. Cardiovasc. Surg.* 108:880–887.

22. O'Rourke, R. A., Peterson, K. L., and Braunwald, N. S. 1973. Postoperative hemodynamic evaluation of a new fabric-covered ball-valve prosthesis. *Circulation* 47,48:74S–79S.

23. Braunwald, N. S. and Bonchek, L. I. 1967. Prevention of thrombus on rigid prosthetic heart valves by the ingrowth of autogenous tissue. *J. Thorac. Cardiovasc. Surg.* 54:630–638.

24. Braunwald, N. S. and Morrow, A. G. 1968. Tissue ingrowth and the rigid heart valve. *J. Thorac. Cardiovasc. Surg.* 56:307–319.

25. Blackstone, E. H., Kirklin, J. W., Pluth, J. R., Turner, M. E., and Parr, G. V. 1977. The performance of the Braunwald-Cutter aortic prosthetic valve. *Ann. Thorac. Surg.* 23:302–318.

26. Schoen, F. J., Goodenough, S. H., Ionescu, M. I., and Braunwald, N. S. 1984. Implications of late morphology of Braunwald-Cutter mitral heart valve prostheses. *J. Thorac. Cardiovasc. Surg.* 88:208–216.

27. Schreck, S., Inderbitzen, R., Chin, H., Wieting, D. W., Smilor, M., Breznock, E., and Pendray, D. 1995. Dynamics of Björk-Shiley convexo-concave mitral valve in sheep. *J. Heart Valve Dis.* 4:21–24.

28. van der Meulen, J. H. P., Steyerberg, E. W., van der Graaf, Y., van Herwerden, L. A., Verbaan, C. J., DeFauw, J. J. A. M. T., and Habbema, J. D. F. 1993. Age threshold for prophylactic replacement of Björk-Shiley convexo-concave heart valves—a clinical and economic evaluation. *Circulation* 88:156–154.

29. Orszulak, T. A., Schaff, H. V., DeSmet, J-M., Danielson, G. K., Pluth, J. R., and Puga, F. J. 1993. Late results of valve replacement with Björk-Shiley valve (1973–1982). *J. Thorac. Cardiovasc. Surg.* 105:302–312.

30. Ericsson, A., Lindblom, D., Semb, G., Huysmans, H. A., Thulin, L. I., Scully, H. E., Bennett, J. G., Ostermeyer, J., and Grunkmeier, G. L. 1992. Strut fracture with Björk-Shiley 70° convexo-concave valve. An international multi-institutional follow-up study. *Eur. J. Cardio-Thorac. Surg.* 6:339–346.

31. de Mol, B. A. J. M., Koornneef, F., and van Gaalen, G. L. 1995. What can be done to improve the safety of heart valves? *Intl. J. Risk Safety Med.* 6:157–168.

32. O'Neill, W. W., Chandler, J. G., Gordon, R. E., Bakalyar, D. M., Abolfathi, A. H., Castellani, M. D., Hirsch, J. L., Wieting, D. W., Bassett, J. S., and Beatty, K. C. 1995. Radiographic detection of strut separations in Björk-Shiley convexo-concave mitral valves. *N. Engl. J. Med.* 333:414–419.

33. Plemons, T. D. and Hovenga, M. 1995. Acoustic classification of the state of artificial heart valves. *J. Acoust. Soc. Am.* 97:2326–2333.

34. Schoen, F. J. and Hobson, C. E. 1985. Anatomic analysis of removed prosthetic heart valves: causes of failure of 33 mechanical valves and 58 bioprostheses, 1980 to 1983. *Hum. Pathol.* 16:549–559.

35. Schoen, F. J., Kujovich, J. L., Webb, C. L., and Levy, R. J. 1987. Chemically determined mineral content of explanted porcine aortic valve bioprostheses: correlation with radiographic assessment of calcification and clinical data. *Circulation* 76:1061–1066.

36. Tomazic, B. B., Edwards, W. D., and Schoen, F. J. 1995. Physico-chemical characterization of natural and bioprosthetic heart valve calcific deposits: implications for prevention. *Ann. Thorac. Surg.* 60:322S–327S.

37. Schoen, F. J., Levy, R. J., Nelson, A. C., Bernhard, W. F., Nashef, A., and Hawley, M. A. 1985. Onset and progression of experimental bioprosthetic heart valve calcification. *Lab. Invest.* 52:523–532.
38. Schoen, F. J., Tsao, W., and Levy, R. J. 1986. Calcification of bovine pericardium used in cardiac valve bioprostheses: implications for the mechanisms of bioprosthetic tissue mineralization. *Am. J. Pathol.* 123: 134–145.
39. Schoen, F. J., Kujovich, J. L., Levy, R. J., and St. John Sutton, M. 1988. Bioprosthetic heart valve pathology: clinicopathologic features of valve failure and pathobiology of calcification. *Cardiovasc. Clin.* 18: 289–317.
40. Levy, R. J., Schoen, F. J., Flowers, W. B., and Staelin, S. T. 1991. Initiation of mineralization in bioprosthetic heart valves: studies of alkaline phosphatase activity and its inhibition by $AlCl_3$ or $FeCl_3$ preincubations. *J. Biomed. Mater. Res.* 25:905–935.
41. Schoen, F. J., Levy, R. J., Hilbert, S. L., and Bianco, R. W. 1992. Antimineralization treatments for bioprosthetic heart valves: assessment of efficacy and safety. *J. Thorac. Cardiovasc. Surg.* 104:1285–1288.
42. Flomenbaum, M. A. and Schoen, F. J. 1993. Effects of fixation backpressure and antimineralization treatment on the morphology of porcine aortic bioprosthetic valves. *J. Thorac. Cardiovasc. Surg.* 105:154–164.
43. Robinson, M. J., Hildner, F. J., and Greenberg, J. J. 1971. Disc variance of Beall mitral valve. *Ann. Thor. Surg.* 11:11–17.
44. Silver, M. D. and Wilson, G. J. 1977. The pathology of wear in the Beall Model 104 heart valve prosthesis. *Circulation* 56:617–622.
45. Schoen, F. J., Titus, J. L., and Lawrie, G. M. 1982. Durability of pyrolytic carbon-containing heart valve prostheses. *J. Biomed. Mater. Res.* 16:559–570.
46. Haubold, A. D. 1994. On the durability of pyrolytic carbon in-vivo. *Med. Prog. Tech.* 20:201–208.
47. Kirklin, J. K., Smith, D., Novick, W., Naftel, D. C., Kirklin, J. W., Pacifico, A. D., Nanda, N. C., Helmcke, F. R., and Bourge, R. C. 1993. Long-term function of cryopreserved aortic homografts. A ten-year study. *J. Thorac. Cardiovasc. Surg.* 106:154–165.
48. Cleveland, D. C., Williams, W. G., Razzouk, A. J., Trusler, G. A., Rebeyka, I. M., Duffy, L., Kan, Z., Coles, J. G., and Freedom, R. M. 1992. Failure of cryopreserved homograft valved conduits in the pulmonary circulation. *Circulation* 86:II-150S–153S.
49. Grunkemeier, G. L. and Bodnar, E. 1994. Comparison of structural valve failure among different "models" of homograft valves. *J. Heart Valve Dis.* 3:556–560.
50. Schoen, F. J., Mitchell, R. N., and Jonas, R. A. 1995. Pathological considerations in cryopreserved allograft heart valves. *J. Heart Valve Dis.* 4:72–75.

12

Polyurethane Pacemaker Leads
The Contribution of Clinical Experience
to the Elucidation of Failure Modes
and Biodegradation Mechanisms

Ken Stokes

1. Introduction

The term "clinical study" can mean many things. Premarket clinical studies are necessary to verify the safety and efficacy of a new device, but they may not be able to detect low-level or long-term complications. No matter how much premarket work one does and no matter how sophisticated the protocols, the only valid proof of long-term reliability is performance in the field; that is, through postmarket surveillance. Even studies on the long-term performance of marketed products may be misleading if not done appropriately. Clinically based postmarket surveillance can reveal the true actuarial survival of a device and the clinical mode of failure. Analysis of returned products may be required to understand the details of failure mechanisms *per se.*

In the case of polyurethane-insulated cardiac pacing leads, we discovered three previously unknown failure mechanisms in marketed products that occurred in spite of state-of-the-art premarket engineering tests and thorough premarket clinical studies. In order to understand how these failures could have occurred, we will briefly review the development and clinical history of the first polyurethane-insulated cardiac pacemaker leads. We will review the discovery of

From: *Clinical Evaluation of Medical Devices: Principles and Case Studies*
Edited by K. B. Witkin Humana Press Inc., Totowa, NJ

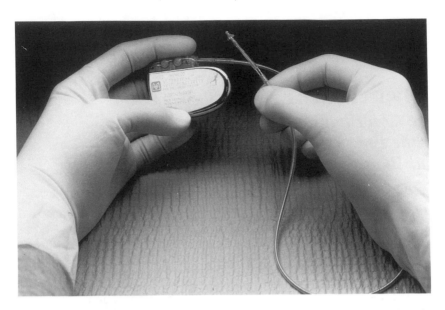

Fig. 1. A typical unipolar single-chamber pulse generator has its electronic circuitry and battery hermetically sealed in a titanium pulse-generator "can." A unipolar lead is attached to the pulse generator. The distal tip of the polyurethane-insulated lead has four pliable "tines" designed to anchor the ring-shaped electrode within the endocardial structures of the heart. Many pacemakers sense and stimulate in both atrial and ventricular chambers, requiring two leads.

these failure mechanisms and how those discoveries changed the way we measure chronic reliability. We will look at the current state-of-the-art interactions between clinical postmarket surveillance and analysis of returned products in the development and monitoring of increasingly reliable implantable cardiac pacemaker leads. However, we must first provide some background about the device itself.

2. The Implantable Cardiac Pacemaker

2.1. The Device

Cardiac pacemakers have two components, a pulse generator and a lead (*see* Fig. 1). The pulse generator includes hermetically sealed circuitry and a battery, with an external connector module. The hermetic container or "can" is usually composed of titanium, whereas the connector modules are typically polyurethane or epoxy. The leads contain one or more metallic conductor coils. The conductors are

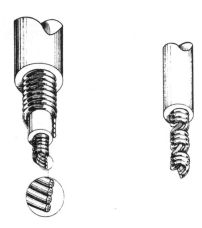

Fig. 2. A unipolar conductor coil is shown on the right. This uses a single, multifilar conductor coil with one layer of insulation. One end of the conductor is connected to a distal electrode, the other to a terminal assembly. A coaxial bipolar conductor is shown on the left. Here the inner insulated conductor runs between a terminal pin and a distal electrode. The outer insulated conductor coil connects with a more proximal second electrode and a ring in the terminal assembly.

insulated with Pellethane 2363-80A, 90A or 55D polyether polyurethane (Dow Chemical), or silicone rubber. The electrodes are usually composed of vitreous carbon, titanium, or platinum/iridium alloy. Unipolar leads have only one insulated conductor (as shown in Fig. 2) and one electrode at the very distal end (*see* Fig. 1). The metallic pulse-generator can is the second electrode in a unipolar pacemaker. Bipolar leads have two conductors (as shown in Fig. 2) and a second electrode 10–28 mm proximal from the tip electrode. The pulse-generator can is not electrically active in a bipolar pacemaker. The distal ends of the leads usually have fixation mechanisms to assure stable contact of the electrode with the endo-or myocardium. These are either "passive" fixation devices, such as the "tines" shown in Figs. 1 and 3, or "active" fixation corkscrews (*see* Fig. 3).

2.2. The Implant Procedure

Transvenous leads are threaded through the venous system to place the electrodes in the heart chamber. This is accomplished either by a venotomy (usually in the cephalic vein) or a subclavian "stick" using a percutaneous introducer.[1, 2] Once the lead tip is in the vein, it is advanced to place the distal tip either in the right ventricle or right

Electrode Retracted

Electrode Extended

Fig. 3. A "cork screw" electrode in its retracted position is shown on the right. During venous insertion and passage, the corkscrew is retracted to protect intravascular tissues. When the lead tip is in position, the conductor coil is rotated to extend the helix, as shown in the center. The corkscrew penetrates the myocardium to provide secure "active" fixation as well as electrical stimulation. A bipolar tined lead tip is shown on the left.

atrium. For dual chamber pacemakers, both an atrial and ventricular lead are placed. After the lead is positioned and tested, it is fixed at the venous insertion site or to the muscle with a suture, with or without an anchoring sleeve, as will be discussed later. The lead terminal(s) are inserted into the pulse-generator connector and are secured with set screws. Then the excess lead is coiled around the pulse generator, which is placed in a subcutaneous or intermuscular pocket.

3. The Polyurethane Lead Story

3.1. Why Polyurethane Leads?

Through the 1970s, the vast majority of transvenous leads were insulated with silicone rubber. Silicone rubber is an excellent implantable material, but its physical properties are limited. It is relatively weak, with low tear strength. In order to provide reasonable protection against mechanical damage (wear, creep, ligature cut-through, and so forth) silicone rubber must be used in relatively thick cross-sections. This was acceptable as long as a single lead was to be threaded through the veins. With the advent of dual-chamber pacemakers,

however, the use of two leads (preferably in one vein) became necessary. Because of a high coefficient of friction (the ratio of frictional
force to the perpendicular force pressing two surfaces together) in
blood (about 0.7), it was difficult to implant two relatively large silicone rubber leads in one vein. The use of two veins tended to increase
postsurgical morbidity and did not necessarily lessen the problem of
one lead dislodging the other when they were manipulated for positioning. Polyester polyurethanes had superior mechanical properties
but were known to be hydrolytically unstable. Polyether polyurethanes, however, were known to be hydrolytically stable, and were
much more durable than silicone rubber.[3] The tear strength of Pellethane 2363-80A (P80A), for example, is about 85 kg/linear cm compared to standard silicone rubber at about 8 kg/linear cm, or the
"high performance" silicones at about 35 kg/linear cm. The mechanical properties of Pellethane 2363 elastomers allowed us to develop
leads that were significantly smaller in diameter, yet less prone to
mechanical damage. Because the coefficient of friction of polyurethane in blood is low (< 0.1), implanting two polyurethane-insulated
leads in one vein was a relatively easy procedure. Thus, the development of polyurethane insulated leads facilitated the use of the more
physiologic dual-chamber pacemakers. In addition, the physical properties of polyurethane allowed us to develop new lead designs that
were not possible with silicone rubber. A good example of this is a
lead with a rotatable terminal pin and conductor coil assembly (Fig. 3).
The lead can be passed easily through the vasculature with the distal
helical "corkscrew" electrode retracted. Then, when the terminal pin
is rotated, the corkscrew electrode emerges from the distal tip to pene-
trate the myocardium, holding the electrode firmly within the tissue.
The lubricity and higher stiffness (or elastic modulus) of polyurethane permits rotation of the conductor coil against the insulation,
whereas in silicone rubber designs, binding prevents extension and
retraction of the helix. The extendible/retractable "corkscrew" design
is very popular today, especially for atrial pacing in dual-chamber
pacemakers.

3.2. The Development and Market Release
of the First Polyurethane Cardiac Leads

In 1975, the first polyether polyurethane-insulated lead was implanted in a human as part of a neurologic stimulator. The develop-

Table 1
Clinically Determined Reoperation Rates for Transvenous Leads

Medtronic model number	Market date	Description	Reoperation rate (%)
Ventricular Leads			
5818	Early 1960s	Straight lead, no fixation	40
6901	Late 1960s	Flanged silicone tip	20
6950	1976	Long silicone tines	12–15
6962	1978	Short silicone tines	5
6972	1980	Polyurethane tines	1
Atrial Leads			
6994	Early 1970s	"J"-shaped silicone	20
6991	1976	Long-tined silicone "J"	7
6991U	1980	Polyurethane tined "J"	2
6957	1980[a]	Transvenous screw-in	1–3

[a] Europe only.

ment of cardiac leads progressed at a slower rate. Several iterations were tested on the bench and in animals before optimized designs were settled on. It was accepted that in canines, cardiac leads became chronically stable well within 12 wk. Longer term testing showed no further significant changes. However, although these data seemed acceptable for evaluating device performance, we were uncomfortable using them for material stability. At that time, there was no history of long-term in vivo materials testing in the literature except for some 8-mo silicone rubber tests by Swanson and Lebeau.[4] Thus, we took the unusual step of conducting a 2-yr evaluation of Pellethane 2363-80A and 55D in rats.[5] Although reversible changes in mechanical properties occurred because of moisture absorption, there was no evidence of instability. The first clinical studies of polyether polyurethane-insulated cardiac pacemaker leads began in Europe in 1977 and in the United States in 1978. Based on excellent animal- and bench-test results, and superior premarket clinical performance, the first polyurethane-insulated cardiac pacemaker lead products were released to US markets in April 1980.[6, 7] As shown in Table 1, the clinical evaluation of the new leads demonstrated significantly improved reoperation rates resulting from lower acute complications.[8] Sales went well, with positive comments about the new lead models excellent performance and ease of implantation.

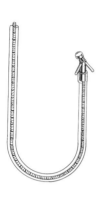
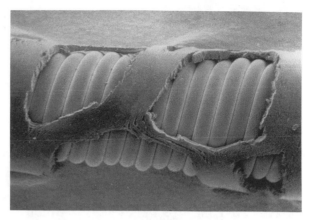

Fig. 4. An illustration of the first polyurethane insulated atrial "J" lead marketed in 1980 is shown on the left. An extreme example of ESC failure at the base of the "J" is shown on the right at about ×30. Note that the edges of the breach match closely. The insulation has cracked and pulled apart as a result of ESC in the presence of unusual tension. No material is missing.

3.3. The Discovery of a New Failure Mechanism, Environmental Stress Cracking

On May 15, 1981, we received a Model 6991U atrial lead that had been explanted from a human after only 5 mo of implantation. The lead had about a 1-in. gap in the insulation at the base of the "J", similar to that shown in Fig. 4. Thorough analysis produced no evidence for chemical degradation of the device.[9,10] For example, we found no changes in the surface or bulk infrared spectra. There was no change in the molecular weight of the sample. However, optical microscopic analysis revealed an interesting occurrence. The edges of the breach, even though separated by about an inch, matched perfectly (*see* Fig. 4). In fact, it was clear that no material was missing, but that the polymer had somehow pulled apart. This lead us to suspect that the insulation had failed because of some kind of stress-cracking mechanism.

The insulation failure was completely unanticipated. We had seen no such cracking in 12-wk canine implants, the 2-yr rat implants, or during premarket clinical studies. Stress cracking mechanisms were not among the known possible degradation mechanisms for polyether polyurethane elastomers. No insulation failures of the first neurologic leads were known, even after 6 yr in service. A thorough review of the literature by us and independent sources found no explanation for this. Thus, we had discovered a previously unknown failure mechanism.

Stress cracking is defined as cracking or crazing of a material in the presence of stress (strain) and a chemical environment.[11] Many forms of stress cracking are known in rigid plastics. For example, polyethylene will crack when bent and exposed to a detergent. Polycarbonate can stress-crack in ethylene oxide because of the residual molded-in stresses. Oxidative stress-cracking is known for many rigid plastics; but, with the exception of natural rubber in the presence of ozone, was not known for elastomers. Indeed, it did not appear to be possible in vivo based on current knowledge. Although we conducted tests on strained polyurethane and many different chemical agents, we could not duplicate the mechanism in vitro. We could not address any chemical component of the mechanism, except that it required exposure to the in vivo mammalian environment. Therefore, the mechanism was labeled environmental stress-cracking (ESC). We did discover, however, that during manufacture the insulation of the Model 6991U lead was occasionally and inadvertently stretched at the approximate point where the failure had occurred in the returned lead. Although we could not control the environmental portion of the mechanism, we could control residual strain in the manufacturing process. Manufacturing techniques were changed to assure that no residual stresses remained in the device as shipped.[12] These changes appeared to be completely effective. Some patches of shallow cracks were still found in the tissue-exposed surfaces of some explanted and returned leads. However, cracks through the insulation to cause clinical failure no longer occurred in the Model 6991U manufactured after the change date. It appeared that the problem had been identified and corrected.

Later in 1981, a few explanted bipolar ventricular leads (Model 6972) were returned with cracks in the insulation around the fixating ligatures (see Fig. 5). Until this point in time, anchoring sleeves had always been supplied separately in the lead packages. The instruction manual indicated that the sleeve should be placed on the lead prior to ligation to prevent damage to the device. It was common clinical practice, however, to ignore the anchoring sleeve and simply ligate the lead directly in the vein. Now that we had identified that the polyurethane insulation was susceptible to a form of stress cracking, it became apparent that ligating the lead directly was no longer accept-

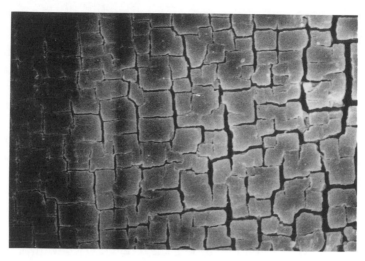

Fig. 5. An electron microscopic view of ESC at a tight ligature at about × 500. The cracks get deeper and wider closer to the ligature (to the right), and decrease in depth and width away from the source of stress (to the left).

able. Therefore, in February 1982, an anchoring sleeve was placed directly on each lead in the factory that could not be ignored and had to be cut off to be removed. The instruction manual was also changed to state that the use of the anchoring sleeve was mandatory. Based on the analysis of returned products, only a small fraction of a percent of the leads that were sold had failed by this or any other mechanism, whereas the next best silicone rubber bipolar ventricular lead had a 5% reoperation rate (refer to Table 1). We were satisfied that the abovementioned changes had solved a problem that had affected very few leads with otherwise superior clinical performance.[13]

3.4. The Development of Accelerated Test Methods for ESC

A more global question was how to determine that design or process changes applied to new products would not cause ESC. How could we prove that the mechanism really was ESC? In our investigation of the stress-cracking mechanism, we recognized several things. ESC cannot occur without strain and the mechanism could not be duplicated in vitro. Obviously, if ESC requires a residual strain then it should depend on processes that increased or reduced residual strain. It was also known that ESC processes required induction periods and critical strains. However, one cannot accelerate ESC in a

thermoplastic by elevating temperature. These polymers are visco-
elastic, which means that they flow or "creep" under load. As temperature increases, the creep rate increases and stresses are relieved.
Therefore, we strained a number of samples over mandrels and implanted them in the subcutis of rabbits. We found that the time to
failure (induction period) varied as a function of the magnitude of
applied strain and the polymer's thermal history. Extrusion conditions and poststrain annealing (a thermal treatment to reduce stress)
were found to have significant effects. Indeed, these properties fit all
of the hallmarks of a stress-cracking phenomenon.[14] The animal
results matched exactly with the findings on explanted and returned
cracked leads.

We settled on a set of standard conditions and developed an accelerated, in vivo test that could be used to evaluate new processes and
new materials. We learned how to optimally stress-relieve the devices
by annealing to prevent ESC failure and incorporated those processes
in evolving next-generation devices.

3.5. The Discovery of Metal Ion Oxidation

In late 1982/early 1983 we received a returned Model 6972 lead
with cracks in the inner insulation. Analysis showed that the polymer
had undergone auto-oxidative degradation and not ESC. Infrared
spectra showed significant chemical changes, including loss of the
aliphatic ether linkages and other changes.[15] Substantial molecular
weight changes were found, which was again unforseen. Auto-oxidation of polyurethanes in the environment is well known, but this process
required several things not believed to be present in vivo. Auto-oxidation is defined by Hawkins as the reaction with oxygen that occurs
between room temperature and about 150°C.[16] It is a free radical
chain reaction that requires the presence of oxygen in reasonable
quantities. However, cardiac pacing leads are implanted in tissues
with very low oxygen tension. In addition, auto-oxidation requires a
catalyst to proceed at clinically significant rates. Photo-oxidation,
for example, requires the presence of both oxygen and certain wave-
lengths of light to initiate and propagate the reaction. The requirement for light is not fulfilled where cardiac pacing leads are implanted.
Thermo-oxidation requires oxygen and a relatively high temperature
to initiate and propagate the reaction. The body produces heat, but

the temperature remains a relatively benign 37 ± 3 °C. Once again, no such phenomena had been seen in preclinical animal implants and were not expected based on the literature. Therefore, a second new and previously unknown phenomenon was presented to us.

3.6. Accelerated Test Methods for Metal Ion Oxidation

As we analyzed the degraded polymer, we began to suspect that metal ions released from the conductor coils may somehow be involved as catalysts.[17] In addition, the medical literature contained new reports revealing the discovery that oxygen free radicals actually *could* be produced in vivo. We discovered from studying the auto-immune disease and pathology literature that the mechanism by which implanted devices become encapsulated in fibrous tissue (known as the foreign body response) involved the release of oxidants. These include hydrogen peroxide (H_2O_2) and oxygen free radicals, such as super oxide anion ($^-O_2$) and hydroxyl radical ($^.OH$).[18, 19] These oxygen free radicals could not possibly affect the inside of the device because of their extreme reactivity. However, H_2O_2 permeates the polyurethane even more rapidly than water to decompose on the metallic conductor wire. A hypothesis evolved that metal ions released from the conductor coils as a result of interactions with H_2O_2 could catalyze auto-oxidative degradation of the polyether portion of the polymer inside the lead.[16] We immersed leads in 3% H_2O_2 and duplicated the mechanism in vitro. Thus, the mechanism was termed metal ion oxidation (MIO).

The first polyurethane bipolar ventricular lead, Model 6972, used a conductor wire made from a composite of silver and MP35N called drawn brazed strand (DBS). MP35N alloy wire (Dupont) has excellent mechanical properties. It is a "super alloy" composed primarily of nickel, cobalt, chromium, and molybdenum. The DBS wire had excellent low electrical resistance and, when coiled, unmatched fatigue-fracture resistance.[20] It had an excellent history when used with earlier silicone-rubber-insulated leads. Based on in vitro testing and analysis of returned products, we determined that DBS wire significantly accelerated MIO in Model 6972, if it was not the cause *per se*. Therefore, that conductor material was removed from the next generation Model 4012, and all future lead designs to be replaced with solid MP35N wire. The next generation product (Model 4012) was expected to have significantly improved performance in all respects, including greatly less, if not the complete elimination of both ESC and MIO. Unfortunately, this is still not the end of the story.

4. Methods of Postmarket Surveillance Used for Cardiac Pacemaker Leads

4.1. Returned Products and Lead Removability

Implanted cardiac pacemakers are often referred to as "permanent" implants. This is a serious misnomer. Pulse generators are battery-operated devices. They have a limited functional longevity, and must be replaced when the battery expires. Although this requires a surgical procedure, it is a relatively simple and risk-free operation performed under local anesthesia. The explanted pulse generators are routinely returned to the original manufacturer for analysis. Compliance is so routine and so good that analysis of returned products provides an excellent means of postmarket surveillance for pulse generators. One can accurately track failure rates and analyze returned devices to learn about failure mechanisms. However, the lead presents other difficulties.

It was well established in the canine model that silicone-rubber leads become encapsulated fully in the heart and vasculature within 12 wk of implant (Fig. 6). Conversely, the polyurethane-insulated leads did not. They typically developed a thin translucent sheath over the distal tip (including the tines and at the ligature site). Encapsulation between the ligature and distal tip was rare. We found that Model 6972 pulled free from the heart with about 750g force once the ligature site was dissected free. Thus, we expected Model 6972 (and its successors) to be chronically removable devices. We were somewhat perplexed to discover that complete removal of a chronic lead from a human often required open-heart surgery. Not all leads were being explanted and returned. Indeed, most of what was being returned was the more readily accessible proximal portion, leaving the distal part of the device in the patient. If the lead failure was distal to the point of separation, then failure could not be verified. In addition, in the mid 1980s we discovered that it was common practice for clinicians to discard the lead, even if it was explanted. We now know that in canines and humans the degree of encapsulation seen at 12 wk persists, even after 2 yr of implantation. It typically takes approx 3–4 yr

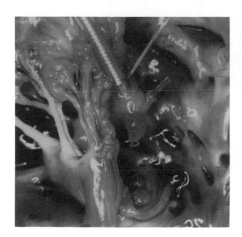

Fig. 6. A typical silicone-rubber-insulated lead after 12 wk implant in a canine is shown on the right. Note the thick encapsulation. A unipolar polyurethane-tined lead after 12 wk implant in a canine is shown on the left. Note the relative freedom from encapsulation except at the distal tip.

for the degree of encapsulation to increase, producing a discontinuous sheath at various points along the lead body. These sheath segments can range from thin transparent collagen, thick opaque white collagen, cartilage, mineralized collagen, and can even contain bone.[3] This clinical finding could not have been predicted even after longer-term animal tests, certainly not within a 2-yr experiment. We needed to validate our perception of chronic lead performance.

4.2. The Development of a Chronic Postmarket Surveillance Study

We contracted three large implanting centers and sent a clinical specialist to visit them. The specialist went through patient records to determine the actual clinical failure rates of various lead models. It was not possible to determine with certainty what caused a clinical failure by this method. For example, oversensing is the situation that occurs when the pulse generator detects something other than the heart's signals, but interprets the artifact as an R-wave and inhibits. If the generator inhibits, this means that it will not emit a stimulus when it should. Such electrical noise can result from various phenomena, including an unstable electrode rubbing against tissue, a loose-set screw in the connector, or metal-to-metal contact within the device. Based on our analysis of returned products, we believed that MIO produced holes in the inner insulation that could allow such metal-to-metal contact. This meant that MIO could be one of several possible causes of oversensing. Thus, the three-center study produced data on clinical failure rates without necessarily identifying the root cause of failure. A preliminary report was issued in the October 1983 issue of *Medtronic News,* which had a circulation of about 38,000 in the medical community.[21] By February 1984, it was clear that the apparent clinical-failure rate was higher than expected, based on the analysis of returned products; it was 7–10% after 3 yr of implantation not < 1%. It must be remembered that there was no industry standard on lead failure rates. We had no knowledge of, nor any way to determine, what our competitor's failure rates were. Nevertheless, this failure rate was not acceptable to us. As a result, we initiated a voluntary advisory on Model 6972.

The only completely acceptable failure rate is 0%, but the reality is that a 0% failure rate for 100% of patients for the remainder of their lifetimes is not possible for implanted cardiac pacemaker leads. As is true of all implanted devices, leads eventually wear out. Therefore, we needed to determine an acceptable failure rate for cardiac pacemaker leads so a trigger point for action could be developed in the unlikely event that one was needed. We asked our customers what would be acceptable to them. We asked questions such as: "How long should a pacemaker lead last?" The almost unanimous initial response was: "For the life of the patient." It was explained that although this was our goal, it was probably not achievable. Our customers typically stated that they wanted the lead to last at least through two pulse generators. At that time, the longevity of a dual chamber pacemaker was commonly about 5 yr or less, which set the time frame at about 10 yr. When asked: "What is the acceptable survival rate for implanted cardiac pacemaker leads after 10 yr?"; the response was typically: "At least 90%." Because 10 yr is a long time to follow these patients, and a long time to wait for a result, we set an interim trigger point at 95% survival at 5 yr. The next question was how to obtain valid data to determine lead survival statistics.

Our initial experience with the three-hospital postmarket clinical study was expanded to presently include 14 large implanting centers. A medical advisory board was set up to review methods, results, and actions. The advisory board meets annually regardless of the study results, but can meet at shorter intervals, if necessary. Thus began the Medtronic "Chronic Lead Study" (CLS), which continues to this day. The results of that study, reviewed by the medical advisory board, are published twice a year.[22]

The CLS requires that each center inform Medtronic whenever a lead complication, patient death, or loss to follow-up occurs. The data analysis assumes that there are no such events at the time of data update unless specifically reported by the center or determined by correlation with returned product analysis. A lead complication is defined as loss of capture (stimulation), loss of sensing, oversensing (detecting a noncardiac signal that inhibits the pulse generator), skeletal muscle stimulation, conductor fracture, insulation breach, or impedance of < 200 Ω. Since this is a chronic study, these complications must be observed at or beyond 1 mo postimplantation. The complication must be resolved by physically modifying, revising (excluding repositioning), replacing, or abandoning the lead. The criteria for a lead complication are summarized in Table 2. These criteria do not enable a lead hardware failure to be differentiated from other clinical events, such as electrode dislodgment, exit block, or concurrent pulse-generator failure presenting as sensing or capture problems. Because the protocol reports clinical and hardware complications, and cannot differentiate between them, there is a likelihood that hardware failures will be overreported. The centers are audited on-site annually to monitor overall compliance with the protocol. In the history of the study, one center has been replaced for noncompliance with the protocol.

The data are analyzed by the Cutler and Eder Life Table method.[23] The actuarial survival curves are reported with standard errors at the leading 3-mo interval. "Survival probability" refers to proper functioning of the device, not the survival of the patient. For example, a survival probability of 98% is a statistical assessment that at the time interval indicated, each patient has a 2% risk of incurring a device

Table 2
Criteria for Lead Complications in the Medtronic Chronic Lead Study[a]

A complication is considered to have occurred in the Chronic Lead Study if both of the following conditions are met.

CONDITION 1: One or more of the following clinical observations beyond 30 d post-implant is reported:

• Failure to capture (stimulate)
• Failure to sense
• Cardiac perforation
• Dislodgment
• Oversensing
• Extracardiac stimulation
• Conductor fracture (observed visually or radiographically)
• Insulation breach exposing the conductor (observed visually)
• Pacing impedance of 200 Ω or less or 3000 Ω or greater

CONDITION 2: One or more of the following clinical actions directly results and is reported:

• Lead abandoned
• Lead explanted
• Lead replaced
• New lead implanted
• Other lead related surgery performed
• Pacemaker mode or polarity reprogrammed to circumvent the problem (i.e., "electrical abandonment")
• Lead use continued, based on medical judgment

[a]Lead positioning is not a qualifying action.

malfunction or complication. An example of the survival curves from the report is shown in Fig. 7.

5. Results From a Clinically Based Postmarket Surveillance Study

5.1. Discovering the Undiscoverable

The Model 4012 lead that succeeded Model 6972 had an actuarial survival rate of about 96%, 5 yr after market release, based on the CLS. Since our trigger point was 95% in 5 yr, this lead was performing acceptably. Although this reflected clinical reality, it did not necessarily tell us what failure mechanisms were affecting 4% of the devices. Looking at returned products that had failed, we discovered that about 25% had ESC breaches in the outer insulation; another 25%

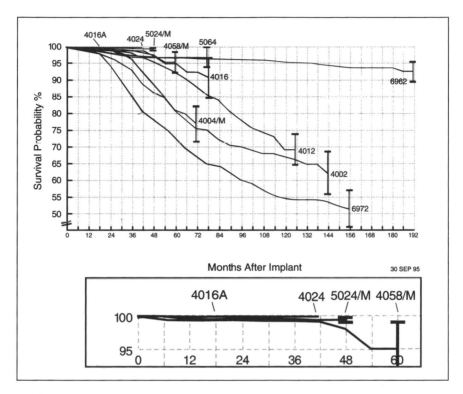

Fig. 7. Actuarial survival curves for the chronic human performance of Medtronic's present and past bipolar ventricular transvenous leads taken from the September 1995 Product Performance Report.

had MIO cracks in the inner insulation. About 50% of the returned failed leads had conductor fracture. Analysis of the returned leads themselves showed that the conductor coils had failed in a crushing mode, which had not been seen previously.[24] Thus, we projected from analysis of returned products that the CLS was showing us 1% ESC failure, 1% MIO failure, and 2% crush failures with 96% failure-free performance after 5 yr of implantation. In comparison, Model 6972 had about 30% ESC and MIO failure in the CLS at 5 yr. Therefore, the improvements made to Model 4012 had been highly effective in reducing ESC and MIO as predicted, although still not to the ultimate goal of 100% survival. We did not anticipate crush failure in the design of Model 4012, because (once again) it was a previously unknown failure mechanism.

5.2. The Crush Fracture Mechanism

Prior to the introduction of Model 6972, leads were typically inserted through the cephalic or jugular vein. A new implant procedure via the subclavian vein had been introduced about the same time that the Model 6972 lead had been in its premarket clinical study. In fact, about 30% of the leads in the premarket clinical study had been implanted by the new method. In this procedure, a needle is inserted percutaneously into the subclavian vein, then a guidewire is inserted into the venous system through the needle. The needle is removed and an introducer inserted over the guidewire. The dilator is removed and the lead inserted into the vein. The introducer and guidewire are removed and discarded. Depending on how the stick was performed, the lead could be positioned between the first rib and clavicle before it entered the vein. In retrospect, we learned that the Model 6972 lead's DBS wire conductor was exceptionally resistant to crush fractures. It would flatten, but not break. Extremely few had been returned for conductor fracture. It was thought that the flat spots found on some returned leads had been caused by instruments at explantation. This conclusion was supported by the presence of marks on the insulation that looked like those made by instruments. DBS wire, however, was found to exacerbate MIO in Model 6972, so it was replaced in Model 4012 with solid MP35N wire that had much better corrosion resistance. Conductor fracture by the well-known mechanism of flex fatigue was relatively rare with multifilar solid wire conductor coils, far less than MIO failures in Model 6972. Analysis of returned products showed us that the fractures occurred at a specific site, 27 $\pm$ 5 cm from the terminal pin. Studies on cadavers demonstrated that the conductor coil could be pressed "out of round" or crushed between the first rib and clavicle if the lead was inserted between the bones before it entered the vein.[25] Tests were done with copper coils and balloon pressure devices.[26] It was found that when the leads were placed between bones before entering the vein, pressures as high as 126 $\pm$ 26 psi could be obtained when the arm was placed in caudal traction. Pressures never exceeded 52 $\pm$ 30 psi when then leads were inserted in the cephalic vein (90° flexion). A survey of all explanted and returned lead records found that all crushed leads had been implanted by the subclavian stick method. In contrast, no returned leads implanted via the cephalic vein had experienced crush fracture. Why were crush fractures not found in preclinical animal studies?

Canines, the generally accepted model for lead studies, have no clavicles. Therefore, it was impossible to discover crush failure in the animal model. Why was it not discovered in premarket clinical tests? The incidence was too low for a premarket clinical study of reasonable length and numbers. We discovered that the marks on the polyurethane insulation were not caused by instruments, but by the pressure of the bones clamping on the device, mimicking instrument damage.

Once the mechanism was understood, modifications to the subclavian stick procedure were made and published.[27] It was recognized that subclavian crush is the result of a surgical procedure, not necessarily a lead design problem. New in vitro test methods were developed to assess crush fracture in new conductor designs. Today, the results of the chronic lead study and analysis of returned products tells us that these have been effective measures.

6. Postmarket Surveillance on Leads Today

So far, the CLS has demonstrated that what we have learned and applied to present-day products has been highly effective. For example, the polyurethane bipolar ventricular transvenous lead, Model 4024, has 100% actuarial survival after 57 mo in the CLS. Its silicone rubber counterpart, Model 5024, has 99.3 ± 0.4% survival after 66 mo. A comparison with Model 4012 at 60 mo shows 92.5 ± 1.4% survival, whereas Model 6972 is at 72.2 ± 3.6%.

A comparison of how one might draw conclusions about chronic lead survival based on two methods is shown in Table 3. Twelve Medtronic lead models are presented based on the March 1997 issue of the Medtronic Product Performance Report. The report contains a table presenting the number of devices sold in the United States since market release, and the number of devices explanted and returned with failure to perform as intended. Data derived from this information is compared in Table 3 to the same lead models followed in the CLS. Based on analysis of returned products, *all 12* lead models are performing well over the 95% trigger point we set for ourselves. One could even claim that Model 6972 has an acceptable 96.4% survival after 13.5 yr. The column representing the results of the CLS, however, shows a very different story. Model 6972 really has only 48.7% survival in 13.5 yr; not 96.4%. Model 4012 has 69.3% survival in 11.5 yr; not 99.2%, as would be surmised from the analysis of returned

Table 3
Comparison of Survival Conclusions Based on Analysis
of Returned Products vs Clinical Postmarket Surveillance

Model number	Years service	Number sold	Returned products % survival	Number in CLS	% Acturial survival, CLS
6962	16.0	70,560	99.7	1418	92.6 ± 2.7
6961	14.0	44,673	99.7	608	89.7 ± 5.5
6972	13.5	43,198	96.4	1253	48.7 ± 5.9
6971	14.3	56,261	99.2	1317	83.9 ± 4.0
4012	11.5	96,901	99.2	2500	69.3 ± 3.8
4011	11.8	64,083	99.8	827	93.7 ± 3.4
4004	7.0	74,481	99.3	1638	68.5 ± 4.5
4003	5.5	39,681	99.9	441	99.5 ± 0.8
4024	4.8	121,647	99.8	721	100.0
5024	5.5	132,395	99.6	5236	99.3 ± 0.4
4057	5.5	11,549	99.6	259	96.4 ± 4.4
4058	6.3	104,540	99.6	1581	95.5 ± 3.2

products. It has now been mandated that all manufacturers must report postmarket surveillance on cardiac pacemaker leads by clinically based processes to the Food and Drug Administration (FDA). So far, however, we see no evidence of clinically based postmarket surveillance being reported to the medical community with the exception of the CLS study.

7. Summary and Conclusions

Polyether polyurethane-insulated cardiac pacemaker leads were introduced to the market with high expectations. They made a significant impact on cardiac pacing, serving many hundreds of thousands of patients well. These leads made dual chamber pacemakers practical to use and facilitated the development of important new designs. Prior to their market release, the first generation of devices were subjected to animal and bench testing that was considered state-of-the-art at the time. Additional testing that exceeded the state-of-the-art was also done, proving that the polyurethane *per se* was biostable. Premarket clinical studies confirmed that these devices were easier to use and had a significantly lower complication rate than their silicone-rubber predecessors. Nonetheless, two previously unknown failure mechanisms, ESC and MIO, were discovered after clinical use, as a result of analysis of returned products (which was the generally accepted means of postmarket surveillance in the pacemaker industry at that time). Improvements were made to existing and new models to reduce ESC and MIO. Even without these improvements, the known chronic clinical failure rates based on analysis of returned products was very low. A clinically based three-center chronic lead study* revealed that the actual failure rate was much higher than was believed because chronic leads were difficult to remove and were not being returned to the manufacturer. The postmarket surveillance results based on this study have been reported to the medical community at least twice a year since 1984.

A third previously unknown lead failure mechanism was discovered in the chronic lead study after escaping detection in animal and pre-market clinical testing. Subclavian crush fracture of transvenous leads escaped detection in animal tests because the mechanism requires bone structures not present in canines or other suitable models. It escaped detection in premarket clinical trials because of its very low incidence (estimated to be about 2% in 5 yr) and appearance (similar to misuse with surgical instruments).

Cardiac-pacemaker-lead postmarket surveillance reporting based on analysis of returned product alone is unacceptable. The FDA has mandated that all manufacturers report on the basis of clinically based chronic lead studies. This may or may not be happening, but so far, only one company reports clinically derived actuarial survival data on its products to the medical community.

References

1. Littleford, P. O., Parsonnet, V., and Spector, D. S. 1979. A subclavian introducer for endocardial electrodes. In *Proceedings of the VIth World Symposium on Cardiac Pacing* (Meere C. ed.). PACESYMP, Montreal, Ch. 14–21.
2. Belott, P. H. 1981. A variation on the introducer technique for unlimited access to the subclavian vein. *PACE Pacing Clin. Electrophysiol.* 4:43–48.
3. Stokes, K., Cobian, K., and Lathrop, T. 1979. Polyurethane insulators, a design approach to small pacing leads. In *Proceedings of the VIth World Symposium on Cardiac Pacing* (Meere C. ed.). PACESYMP, Montreal, Ch. 28–32.

* The first CLS study included three centers. Later, the study was expanded to 11, and then 14, centers. Additional centers are being identified in Europe and Asia.

4. Swanson, J. W. and Lebeau, J. E. 1974. The effect of implantation on the physical properties of silicone rubber. *J. Biomed. Mater. Res.* 8: 357–367.

5. Stokes, K. and Cobian, K. 1982. Polyether polyurethanes for implantable pacemaker leads. *Biomaterials* 3:225–231.

6. Stephenson, N. L. 1980. Synopsis of clinical report on the Spectraflex models 6971/71 transvenous leads. *Medtronic News* X(3):16.

7. Stephenson, N. L. 1980. Synopsis of clinical report on models 6990U/ 6991U atrial J leads. *Medtronic News* X(2):10.

8. Stokes, K. and Stephenson, N. L. 1982. The implantable cardiac pacing lead—just a simple wire? In *Modern Cardiac Pacing* (Barold S. and Mugica J. eds.). Futura, Mount Kisco, pp. 365–416.

9. Stokes, K. B. 1982. The long-term biostability of polyurethane leads. *Stimucoeur* 10: 205–212.

10. Timmis, G. C., Gordon, S., Westveer, D., Martin, R. O., and Stokes, K. 1983. Polyurethane as a pacemaker lead insulator. In *Cardiac Pacing* (Steinbach K. ed.). Steinkopff Verlag, Darmstadt, pp. 303–310.

11. Whittington, L. R. 1968. *Whittingtons Dictionary of Plastics.* Technomic, Stanford, pp. 90.

12. Stokes, K. 1984. The biostability of polyurethane leads. In *Modern Cardiac Pacing* (Barold S. ed.). Futura, Mount Kisco, pp. 173–198.

13. Stokes, K. B. 1984. Environmental stress cracking in implanted polyether polyurethanes. In *Polyurethanes in Biomedical Engineering* (Planck H. Egbers G. and Syré, I. eds.). Elsevier, Amsterdam, pp. 243–255.

14. Stokes, K., Urbanski, P., and Cobian, K. 1987. New test methods for the evaluation of stress cracking and metal catalyzed oxidation in implanted polymers. In *Polyurethanes in Biomedical Engineering II* (Planck H., Egbers, G., and Syré, I. eds.). Elsevier, Amsterdam, pp. 109–128.

15. Stokes, K. B., Berthelsen, W. A., and Davis, M. W. 1985. Metal catalyzed oxidative degradation of implanted polyurethane devices. Proceedings of the ACS, Division of Polymeric Materials Science and Engineering, 53:6–10.

16. Hawkins, W. L. 1972. *Polymer Stabilization.* Wiley-Interscience, New York.

17. Stokes, K., Urbanski, P., and Upton, J. 1990. The *in vivo* auto-oxidation of polyether polyurethanes by metal ions. *J. Biomater. Sci. Polym. Ed.* 1:207–230.

18. Anderson, J. A. 1988. Inflammatory responses to implants. *ASAIO* II:101–107.

19. Sybille, Y. and Reynolds, H. V. 1990. Macrophages and polymorphonuclear neutrophils in lung defense and injury. *Am. Rev. Respire Dis.* 141:471–501.

20. Upton, J. E. 1979. New pacing lead conductors. In *Proceedings of the VIth World Symposium on Cardiac Pacing* (Meere C. ed.). PACESYMP, Montreal, Ch. 29–36.

21. Helland, J. 1983. Pacemaker lead complications: clinical significance and patient management. *Medtronic News* XIII:8.
22. Medtronic, Inc. 1995. *Medtronic Product Performance Report* UC-9300222eEN, September, Minneapolis, MN.
23. Cutler, F. and Eder, F. 1958. Maximum utilization of the life table method in analysis of survival. *J. Chronic Dis.* 699–712.
24. Stokes, K., Staffanson, D., Lessar, J., and Sahni, A. 1987. A possible new complication of subclavian stick: conductor fracture. In *VIII World Symposium on Cardiac Pacing and Electrophysiology.* Jerusalem: *PACESYMP,* 10(3), Pt. II, 748 (Abst. 476).
25. Fink, A., Jacobs, D. M., Miller, R. P., Anderson, W. R., and Bubrick, M. P. 1992. Anatomic evaluation of pacemaker lead compression. *PACE Pacing Clin. Electrophysiol.* 15:510.
26. Jacobs, D. M., Fink, A. S., Miller, R. P., Anderson, W. R., McVenes, R. D., Lessar, J. F., Cobian, K. E., Staffanson, D. B., Upton, J. E., and Bubrick, M. P. 1993. Anatomical and morphological evaluation of pacemaker lead compression. *PACE Pacing Clin. Electrophysiol.* 16:434,444.
27. Byrd, C. L. 1992. Safe introducer technique for pacemaker lead implantation. *PACE Pacing Clin. Electrophysiol.* 15:262–267.

Glossary

Analytical sensitivity: The lower limit of detection of an analyte by an assay system.

Attributable risk: The proportion of disease incidence or other outcome in exposed individuals that can be attributed to a specific exposure. This measure is derived by subtracting the rate of the outcome in those who are not exposed from the rate in those who are exposed.

Bias: Refers to a trend in the collection, analysis, interpretation, publication, or review of data that can lead to conclusions that are systematically different from truth. There are many sources of bias; some of the most significant include selection, recall, ascertainment, and misclassification biases, as well as confounding.

Case-control study: Observational epidemiologic study in which persons with an outcome of interest or disease (cases) are compared with a group of persons without the outcome or disease (controls). Exposure characteristics or other risk factors are compared between cases and controls. A case-control study can be called "retrospective" because it starts after the onset of disease and looks back to the postulated causal factors.

Case report (case series): A report of a disease or health outcome in a single patient or series of patients. Case reports are descriptive in nature; although they can suggest areas for further study they cannot be used to draw conclusions about causation, primarily because the influence of other factors or pre-existing disease cannot be eliminated as a cause of the observed symptoms.

From: *Clinical Evaluation of Medical Devices: Principles and Case Studies*
Edited by K. B. Witkin Humana Press Inc., Totowa, NJ

Clinical follow-up study: A study in which individuals who have been exposed to a risk factor or condition (such as device implantation) are followed to assess the outcome of the exposure. Typically, there is no comparison group in a clinical follow-up study.

Cohort study: A cohort study typically follows a group of exposed (treated) and nonexposed (untreated) individuals either forward in time (concurrent design) or retrospectively (nonconcurrent design, that is, from a specific point in time in the past to the present), to observe who develops a disease or outcome.

Concurrent cohort study: *See* Cohort study.

Confounding: Confounding can arise if the association between exposure to a factor of interest and consequent development of an outcome is distorted by an additional variable **(confounder)** that is itself associated both with the factor and the outcome of interest.

Endpoint: For medical devices, the last follow-up interval for a clinical study after implantation of the device. This total study duration should be long enough for the selected parametric measurements to reveal either success or failure of the device. Historically, clinical trials of orthopedic devices have been arbitrarily required to last for a minimum of 24 mo.

Epidemiology: The study of patterns of disease occurrence in human populations and the factors that influence these patterns.

Expected range of analyte: The range of analyte concentrations within which a disease-free individual would be expected to fall.

Experimental study: A study in which conditions are under the direct control of the investigator. In this type of study, a population is selected for a planned trial of a regimen whose effects are measured by comparing the outcome of the regimen in the experimental group with the outcome of another regimen in a control or appropriate comparison population.

Functional sensitivity: The lowest analyte concentration an assay can detect with reasonable precision and reliability.

Hazard Rates: An idealized or theoretical rate for a small time interval; considered an instantaneous risk of failure.

Health Status Assessment: Measurement or evaluation of a person's or patient's health, as determined by self-reported responses to a questionnaire that can contain biological, functional, mental, social, emotional, and other indicators.

Historical controls: A historical control group is a group of individuals separated in time or place from the study population, but received no intervention or a different intervention for the same disease or condition. A historical control group is also called a nonconcurrent control group.

Institutional Review Board (IRB): A committee that has been formally designated by a medical institution for the purpose of review and approval of biomedical research involving human subjects, in order to protect their rights and welfare.

In Vitro Diagnostic Devices (IVD): Medical device products that are intended for use in the collection, preparation, and examination of specimens taken from the human body; include laboratory instruments, reagents, and assays.

Linearity: The measurement of a laboratory system's ability to precisely detect analyte concentrations over the entire assay range.

Method comparison: A clinical study designed to compare a new or test method with a cleared, approved, or licensed product.

National Committee for Clinical Laboratory Standards (NCCLS): Provides consensus guidelines for evaluation of laboratory systems, assays, and reagents.

Negative Predictive Value (NPV): The probability that a clinical sample testing negative with a particular assay really does not contain any of the analyte being measured.

Nonconcurrent cohort study: *See* Cohort study.

Nonexperimental study: A nonexperimental study, also known as an observational study, is a study in which there is no intervention by the researcher. Nature is allowed to take its course; changes or differences in one characteristic are studied in relation to changes or differences in other characteristics.

Observational study: *See* Nonexperimental study.

Odds ratio (OR): The ratio of two odds. In a case control study, this is the ratio of the odds in favor of getting the disease, if exposed, to the odds in favor of getting the disease if not exposed.

Outcome: Something that follows as a consequence of some antecedent action or event. In the context of clinical studies of medical devices, an event or measure observed or recorded for a particular person or treatment unit in a trial during or following implantation or use of the device that is used to study the safety and effectiveness of the medical device. For medical devices, clinical outcome is influenced by both the product or device under evaluation, as well as the skill and discretion of the user.

Outcome domains: The conceptual areas that are encompassed by describing the state of an individual or individuals at given time periods post-illness and/or device, drug, procedure, or other treatment intervention.

Outcome measurements: A measure of patient status following exposure to a preventive or therapeutic intervention. Outcomes of interest in medical device research can include patient satisfaction, quality of life, change in clinical function, pain relief, cost of treatment, adverse effects, or longevity. Outcome measurements must be accompanied by validated parametric tools for maximum utility.

Parametric observation: A standard recognized tool used to determine efficacy of treatment. In clinical studies, parametric observations can include functional parameters, such as assessment of pain, range-of-motion (or measurable clinical function), and activities of daily living; laboratory or radiographic evaluations; and the reporting of complications.

Patient-centered outcome: An outcome descriptor that focuses on the patient's status or results rather than the properties of the intervention.

Pilot studies: Limited investigations of a device that are intended to provide data on preliminary evaluations of safety and performance, i.e., the feasibility of the device for diagnostic or therapeutic clinical use. Also called feasibility studies.

Positive Predictive Value (PPV): The probability that a clinical sample testing positive with a particular assay really does contain the analyte being measured.

Postmarket Surveillance: Any procedure or system implemented on or after approval of a device or drug for a given indication or use, designed to provide ongoing information on the use of the device or drug for that indication and on its side effects. The surveillance usually involves survey and observational techniques and is usually initiated in response to needs expressed by the Food and Drug Administration, the manufacturer, or some other group for added information concerning use or safety.

Precision: The ability of a laboratory assay to obtain the same result when repeatedly measuring an analyte in a clinical specimen.

Qualitative assays: Those laboratory assays that simply detect the presence of an analyte in a clinical specimen.

Quality of Life Measurement: An individual's overall satisfaction with life, and one's general sense of personal well-being.

Quantitative assays: Those laboratory assays that are able to measure the concentration of an analyte in a clinical specimen.

Randomized clinical trial: A study whereby subjects are randomly allocated to an intervention. The intervention group is compared on the outcome(s) of interest to a control group that does not receive the intervention.

Relative risk (RR): The ratio of the risk of disease or death among the exposed to the risk among the unexposed.

Retrospective study: Generally defined as a study in which the logic of the design leads from effect to cause. The investigator begins with people who have the outcome or disease of interest, and then examines exposures or experiences that they had prior to the outcome. Often such studies rely on data that were collected before the study was designed; sometimes, however, new data collection is necessary.

Significant Risk (SR): An investigational implanted device that presents a potential for serious risk to the health, safety, or welfare of a subject. SR devices can be used in supporting or sustaining human life; or may be of substantial importance in diagnosing, curing, mitigating or treating disease, or otherwise prevent impairment of human health; or otherwise present a potential for serious risk to the health, safety, or welfare of a subject.

Utility Measures: Normative models for individual decision making under uncertainty or numbers that represent the strength of an individual's preferences for particular outcomes under uncertain conditions.

Index